Organizational Mastery with Integrated Management Systems: Controlling the Dragon

Michael T. Noble

Organizational Mastery with Integrated Management Systems: Controlling the Dragon

A Manager's Tool Box for Enhancing Process Quality and Environmental Health and Safety (QEH&S)

Michael T. Noble

A John Wiley & Sons, Inc. Publication

New York • Chichester • Weinheim • Brisbane • Singapore • Toronto

This book is printed on acid-free paper. ♾

Published simultaneously in Canada.

For ordering and customer service, call 1-800-CALL-WILEY.

Library of Congress Cataloging-in-Publication Data:

Noble, Michael T.
Organizational mastery with integrated management systems : controlling the dragon / Michael T. Noble.
p. cm.
"A manager's tool box for enhancing process Quality and Environmental Health and Safety (QEH&S)."
"A Wiley-Interscience publication."
Includes index.
ISBN 0-471-38928-5
1. Industrial safety—Management. 2. Quality assurance—Management. I. Title.

T55 N53 2000
658.3'82–dc21 00-020685

Printed in the United States of America.

10 9 8 7 6 5 4 3 2 1

Contents

Forewords **xi**

Joe Cascio, Global Environment & Technology Foundation; Chair, U.S. Technical Advisory Group for ISO 14000 *xii*

Steve Levine, Ph.D., CIH, University of Michigan, President-Elect of AIHA *xiii*

Kathy A. Seabrook, CSP, RSP (UK), President, Global Solutions, Inc., Vice President, Practices and Standards, American Society of Safety Engineers *xiii*

Frank Mastrioanni, SrVP Marsh, Inc. *xiv*

Robert C. Wilson, President of IQuES, a leading ANSI-RAB accredited ISO 14000 training and consulting company *xiv*

Vic Johnson, Senior Vice President, Harding Lawson Associates *xiv*

James D. Griffin, MPH, CIH, Manager, World Wide Facilities, General Motors Corporation *xv*

Mark Kline, Regional Well-Being Manager, IBM Corp. *xvi*

Huw Andrews, Bsc(Hons), FIOSH RSP, Director, Health, Safety, and Environment, EEF South, United Kingdom *xvi*

About the Author **xvii**

Acknowledgements **xix**

Introduction **xxi**

Chapter 1 Controlling the Dragon — The Integration of Management Systems **1**

ISO 14000/9000 Harmonized Quality and Environmental Health and Safety (QEH&S) Management Systems 1

ISO 9001/14001 Harmonized Management Systems 8
Summary 22
Action Plan Considerations 23

Chapter 2 Building Safer, More Productive Jobs! 27

Management 32
Estimating Indirect Cost 35
Safety is Productivity 38
Controlling Risk and Cost 39
Action Plan Considerations 43

Chapter 3 Management Commitment: Policy and Planning 44

Commitment and Policy 48
Communication 49
Responsibility and Authority to Act 50
Summary of Key QEH&S Management System Policy Requirements 51
Management Job Influences 51
Objectives and Targets 53
Plan Considerations 55

Chapter 4 Plan Implementation 57

Implementation and Operations 58
Human, Physical, and Financial Resources 63
Integration 63
Employee Involvement and Consultation 64
Communications 64
Procedures: Process Control Instructions (PCIs) or Standard Operating Procedures (SOPs) 66
Management and Control of Contractors and Vendors 68
Emergency Preparedness and Contingency Planning 69
QEH&S Management System Document Control and Record-Keeping Processes 69
Process Risk Analysis of Process Inputs 74

Chapter 5 Risk Assessment Process 80

Evaluating Business and Organizational Quality and Environmental Health and Safety (QEH&S) Risk 81
Cost–Benefit Analysis of Risk Reduction Measures 83
Alternative Risk Management Strategies 85

Risk Financing 88
Alternative Risk Financing 89
Risk Assessment Process Summary 95

Chapter 6 Risk and Operational Control 113

Hazard Risk Identification 115
Purchasing 118
Process Improvement — Risk Reduction 123
Training 126

Chapter 7 Measurement and Evaluation 133

Goals of Measurement and Evaluation System 138
Performance or Behavioral Safety Management 141
Inspection and Evaluation (Monitoring) 145
Job–Task Observations and Analysis 154

Chapter 8 Nonconformance and Incident Investigation 187

Process Failure/Incident Investigation Policy 188
Why Investigate Accidents or Process Nonconformances, Including Near-Misses or Incidents? 189
Customer Complaint/Rework and Accident, Incident, and Illness Data as QEH&S Performance Indicators 190
Process Failure/Incident Analysis 191
The Carelessness Myth 191
Key Assessment Considerations 192
Recommendations and Follow-Up 194
Process Nonconformance and Incident Investigation Principles 194
Learning From and Communicating Investigation Results 196
Nonconformance Accident–Illness Investigation Process Policy 198
Program Synergism/Conclusion 198
Action Plan 199

Chapter 9 Management Review and Continuous Improvement 213

Corrective and Preventive Action 214
Program Audit Attributes 215
Audits, Measurements, Inspections, or Testing Activities 216

Internal Audits of QEH&S Systems 217
The Action Plan 219

Chapter 10 Conclusion 221
Changing Your Business 233

Appendix I: Audit Attributes 235
Policy and Management Commitment 235
Responsibility, Accountability, and Authority 237
Management and Staff Job Descriptions 238
Quality System 238
Implementation and Operation 239
Legal and Other Requirements 239
Objectives and Targets 239
Communications 240
Risk Assessment Process 241
Measurement and Evaluation 242
Audits, Measurements, Inspections, or Testing Activities 243
Documentation and Records 246
Corrective and Preventive Actions 249
Risk and Operational Control 250
Design, Process and Engineering Control 250
Training 254
Purchasing 256
Handling, Storage and Packaging of Materials, Including HM/HW 256
Management of Contractors and Suppliers 260
Emergency-Contingency Preparation 261
Maintenance/Servicing 262
Nonconformance and Accident Investigation 262
Management Review and Continuous Improvement 263
Office Quality Assessment Nonconformance Report Tracking Form 265

Appendix II: Miscellaneous Environmental Health and Safety Program Audit Attributes and Technical Check Sheets 266
Medical and Occupational Health 267
Electrical Safety 269

Permit System: Hot Work and Confined Space Entry 272
Lock-Out/Tag-Out Procedures 273
Release from Lock-Out or Tag-Out 274
Life–Fire Safety 277
Respiratory Protection and Personnel Protective Equipment 278
Tools–Equipment, Including Maintenance 279
Inspection Checklist for Pneumatic Tools 279
Inspection Checklist for Hydraulic Power Tools 279
Inspection Checklist for Powder-Activated Tools 279
Hoist and Auxiliary Equipment 279
Abrasive Wheel Equipment-Grinders 280
Hazardous Materials and Hazardous Waste Requirement Checklist 280
Manual Material Handling 283
Miscellaneous Safety Concerns 284

Appendix III: Who is Lincoln Electric and What Is the Lincoln Incentive Management System? 286

Appendix IV: Comparison of Safety, Environmental and Quality Management Systems 290

Appendix V: SAIC: The Largest Employee Owned Company in the World, #347 in the Fortune 500 Listing 293

References 295

Index 298

Forewords

In 1996, the opportunity to develop yet another international ISO standard for occupational health and safety was turned down by the overwhelming consensus of representatives from industry, labor, and various governments from both developed and developing countries. There were many contributing reasons for this unusual and unexpected consensus, among which was the realization that organizations cannot and should not be expected to implement multiple, overlapping, redundant, and possibly conflicting management systems. That year, ISO 9000 was already well-established throughout the world and ISO 14000 had just been released after a relatively short three-year development effort in ISO Technical Committee 207. There was very little appetite at that point for yet another set of specifications that would inevitably lead to a new layer of system procedures, programs, and metrics for organizations to implement and seek certification. The belief at that time was that these standards were additive and that they resulted in multi-layered systems within an organization.

In the intervening years, that view has been overtaken by the realization that an organization should really have only one management system — not a stack of them — and that the specifications in ISO standards are simply elements that can and should be integrated into the existing organizational system. ISO 9001 and ISO 14001, and all such "system standards", are now more often considered as sources of system elements that address a specific part of an organization's structure and dynamics. Once integrated, these elements become components of the overall system, undifferentiated and totally congruent with the underlying principles of employee involvement, management leadership, process consistency, prevention, and continual improvement. These systems are also expected to include common organizational elements for training, corrective and preventive actions, and communications, document control, record keeping, system audits, and periodic reviews by top management.

Mr. Noble has made the strongest case yet for the integration of disparate management systems in the quality, environmental, health, and safety areas. This volume presents a very compelling case for such integration, while usefully providing an abundance of tools, methods, and checklists that the prospective users can apply to their own systems. In an academic setting, this volume could effectively serve to provide a pragmatic view of management systems theory to students of safety, environmental, or risk management disciplines. This text may also engage the parties and breathe new life in the ongoing debates on the desirability of an OH&S standard and on the best way to promote system integration.

—Joe Cascio, Global Environment & Technology Foundation; Chair, U.S. Technical Advisory Group for ISO 14000

For the person who is a professional in the field of risk evaluation and management, this volume will prove to be a useful source of information. This book has a richness of figures, "how-to" diagrams, and process analysis models that complement the explicit instructions to auditors. It has a nice focus on OHS management systems, which are important today and will be ever more important in the future.

The compelling case for an integrated occupational hygiene, safety, and environmental management system integral to the quality system of an organization is made in this book. Because of its efficiency and effectiveness, this integrated system will work. Even if we ignore the compelling ethical arguments in favor of using all available effective tools in our arsenal to protect workers domestically and globally, this book explains why clear-seeing organizations will implement such an approach because this approach represents good business.

ISO 10441 as a pure environmental management system without OHS aspects integral to its philosophy and practice simply was, is, and will be illogical. Even in the meeting in Chicago during which "international ISO standard for OHS was turned down by the overwhelming consensus of representatives from industry, labor, and various governments from both developed and developing countries", this point was already moot. At just about the time of that meeting, an article was published in the technical press entitled "IBM spells SAFETY I-S-O". Indeed, a representative of one of the largest companies in the world stood up in that Chicago meeting and said, "Whom do we think we are kidding? We are either already doing this [integrated E&OHS MS], or are planning to do it because it makes sense!"

This book explains why that is so, and how to achieve such an integration. Let us hope that practitioners and policy makers alike will embrace this idea. Indeed, there are many such efforts in progress. The British Standards Institute has its BS 8800 and BS 18001 products; the American Industrial Hygiene Association has its ISO 9001-based OHS MS and its ANSI Z-10 committee; and the ILO has its landmark report published in 1999 surveying and analyzing the world's

OHS management system standards. These groups, and Michael Noble with this book, are leading the way. One result will be that this practice becomes accepted worldwide, as well as a clearly integrated part of ISO 14001:200X, and thereby part of the world's contract specification systems.

—Steve Levine, Ph.D., CIH, University of Michigan, President-Elect of AIHA

In *Organizational Mastery with Integrated Management Systems*, Michael Noble has achieved a work for leaders who oversee the strategic management of quality, environmental, health and safety (QEH&S) issues within their organizations. It is a key process risk management tool to integrate QEH&S policy, planning, implementation, measurement & evaluation, management review and continuous improvement strategies.

To maintain a competitive advantage in the global economy, organizations are looking toward third party verification of their product quality, environmentally sound manufacturing processes and worker health and safety. The International Standards Organization is internationally recognized by business as the standards setting body for Quality Management Systems (ISO 9000) and Environmental Management Systems (ISO 14000). The integrated management systems approach presented in this work is aligned with these internationally recognized standards.

Organizational Mastery with Integrated Management Systems sets QEH&S in the language of business, not the typical quality, environmental, health or safety technical jargon that stands as a barrier to organizational understanding and implementation of business solutions. Of particular note are the real world examples which provide a "how to" approach to integrated management systems. This approach includes case studies and "hands-on" tools such as a Critical Process Observation Checklist and Job-Task Observation Report.

The influence of this book on the future of quality, environmental, health and safety management will be considerable. You will find this may be the only book you will need to assist you in meeting the challenges of an ever changing global economy, workplace, workforce and regulatory environment.

—Kathy A. Seabrook, CSP, RSP (UK), President, Global Solutions, Inc., Vice President, Practices and Standards, American Society of Safety Engineers

Mike Noble understands that safety is often not well-managed within an organization, and that there is a better framework for safety management. This book will be useful to several levels of business people. It will be valuable to management consultants, safety consultants, corporate safety directors, risk managers, financial officers, manufacturing mangers, and production managers. The topics that Mike discusses are not new, but he has arranged them in a new and very readable way. I have discussed these topics, in one form or another, with my clients in recent years. It was good to see them all in one book.

Safety consultants and management consultants will use the information they learn to spur their clients to move to the next level of corporate control of their resources. Risk managers will use the information as a guideline to move forward with their management and with their chief financial officers (CFOs).

Management and CFOs will find the book easy-to-read. They will understand why their safety and management consultants are suggesting that certain processes be adopted. I like the way Mike has concluded each chapter with an "Action Plan". Each level of reader will get something different from each chapter and will develop an action plan and strategy based on his or her insights from the text. Covering all this material, anyone wishing to implement Mike's suggestions will have the same goal, but each will have a unique understanding of his or her role in the process.

I have read other safety texts published by John Wiley and Sons, Inc. "Controlling the Dragon" is one of the books that will make a difference for me in terms of improving my thought process, in presenting ideas to clients, and enhancing my consulting ability.

—Frank Mastrioanni, SrVP Marsh, Inc.

With the rapid growth of ISO 14000, integration of management systems into a complete operating system is a key objective of many organizations. Mr. Noble's book deals with this timely subject in a comprehensive and lively manner. He not only covers the specific items relative to an ISO 9000 QMS and an ISO 14000 EMS but also is able to show the relationships in a way that makes good common sense.

Beyond Mr. Noble's coverage of ISO 9000 and ISO 14000, he breaks new ground in integrating concepts of health and safety (H&S) into his coverage of an integrated management system. A prospective "ISO 16000" for H&S was tabled by ISO, but the need for integrating H&S issues into the corporate management system nonetheless remains. Indeed, a significant multi-national client of ours has specifically requested incorporating H&S procedures into its ISO 14000 and QS-9000 management system. Mr. Noble's book provides in-depth guidance on doing just that.

I think Mr. Noble's book could become a standard treatise on the subject, and I look forward to its publication.

—Robert C. Wilson, President of IQuES, a leading ANSI-RAB accredited ISO 14000 training and consulting company

The United States, in general, has historically provided high-quality goods and services. Unfortunately, U.S. firms have used many different systems in providing these quality goods and services. Each of these systems has strong vocal advocates, and collectively they result in a lot of "quality noise" in the marketplace. This result, in turn, makes it difficult for practicing professionals to discern what really is important and what isn't.

As a senior officer of a publicly held engineering firm for the past 19 years, I have often been confused with all the professional "quality noise" in the marketplace. The real compelling value of Mr. Noble's work is that he takes this large mass of quality information and boils it down into something that one can readily understand. I enjoyed reading the book and I think significant numbers of other professionals will also.

—Vic Johnson, Senior Vice President, Harding Lawson Associates

It has been a professional highlight for me to be asked to review and comment on Mr. Noble's book "Controlling the Dragon". As a member of the Occupational Safety & Health field since 1973, I have had many profitable experiences but none have equaled the opportunity to contribute to such a comprehensive body of work.

The high watermarks of my career include positions of leadership, management, strategic planning, and program directing requiring knowledge of key literature that defines the state-of-the-art in this profession. Unfortunately, all of my quality program experience has been manufacturing-oriented without regard to service fields such as environmental health and safety (EH&S). Similarly, as a developer of national and global programs, I have continually struggled to provide direction to EH&S resources in the effort of quality systems and programming. These endeavors had led to the realization of the unfortunate fact that no one until Mr. Noble has approached the ES&H field with as comprehensive a volume of literature designed to make a major impact on program and system quality.

Mr. Noble's expertise in the EH&S field has allowed him to organize this book and present the "quality" message in a way that industry and the EH&S field will benefit considerably. In my view, Mr. Noble earns his credibility throughout the book starting with the introduction, as he details the historical perspective of "quality management", to the conclusion, wherein all levels of management are provided with a road map to success. In addition, the full text contains a richly filled toolbox to assist management in the design of policies, strategic planning, business improvement scenarios, quality EH&S program development, audits, cost–benefit analyses, and training.

The added bonus many multi-national companies will discover is that the concepts, principles, tools, plans, programs, processes, templates, forms, and overall philosophies in the book are global in nature and can be applied to any system worldwide. This information will better prepare companies for the ISO certification that will be required to compete in the global market.

In summary, I support Mr. Noble's concept of integrating EH&S into a company's "quality management system".

—James D. Griffin, MPH, CIH, Manager, World Wide Facilities, General Motors Corporation

Many multinational companies are experiencing fast paced business growth and global expansion into developing markets and countries. This expansion challenges the traditional means of providing occupational health and safety (OHS) services to their employees. Within these countries there may be significant variations in the scope, content and effectiveness of employee safety and health standards and regulations. Professionals along with line management are challenged to fine new ways to communicate and apply internal standards and processes within these variable work environments.

In response, many companies have turned toward management system standards as a means to develop and deploy occupational health and safety

(OHS) processes in highly complex business environments. These management systems are generally modeled from the ISO 14001 framework. Although there is no international ISO standard for occupational health and safety, companies are demonstrating leadership in this area by developing and institutionalizing internal management systems as a way of integrating OHS solutions into business strategies an operations.

For those companies who are incorporating OHS concepts into new or existing management systems, "Controlling the Dragon" is an exceptional guidance and reference document. The concept of utilizing an integrated management system approach as a methodology to develop these systems is clearly the way to go. The book provides clear guidance and will challenge professionals to focus on the entire management system rather than the concentric task at hand. In addition, it provides management system coordinators and focal points with risk analysis concepts, operational control techniques and measurement tools necessary to ensure effective maintenance and development of quality management systems.

The book will also help companies identify significant aspects, objectives and targets which are essential in any quality management system. Clearly, Mr. Noble's book is timely and will be a valuable tool to professionals for years to come.

—Mark Kline, Regional Well-Being Manager, IBM Corp.

The European Union continues to reflect—and in some cases, lead—the globalization of accepted corporate practices; benchmarks; and reflected accountabilities in relation to the management of health, safety, quality, and environment. Compliance with ISO 9001 and 14001 has gone beyond mere certification to become established values, whilst management systems in relation to health and safety have been an explicit legal requirement since 1992.

On the eve of a certified management standard for health and safety (in the United Kingdom, at least), Mike Noble's book comes at a time when the impetus in Europe for a truly integrated system to prevent loss has never been greater. His book will fuel the debate and go beyond the mere academic—offering practical guidance and methodologies for those wishing to rise to the challenge.

As such, *Controlling the Dragon* is sure to become an invaluable source for practitioners and students alike at a time when industry is not only seeking to lend transparency to its corporate practices but also to ensure the prevention of accidents and to reduce loss and secure the environment for future generations.

—Huw Andrews, Bsc(Hons), FIOSH RSP, Director, Health, Safety, and Environment, EEF South, United Kingdom

About the Author

Mike Noble has 23 years of experience as an environmental health and safety risk manager and has been a consultant since 1989. His senior management experience includes international health and safety consulting with a global insurance company, and the startup of a public environmental consulting company. He has done a lot of work in operations management as well, including strategic planning, marketing and business development, product/service development, budget and personnel management, labor relations, training, and quality and project management.

Mr. Noble also has a great deal of technical experience—developing and teaching courses for Southern Illinois University, Liberty Mutual, and Liberty International Risk Services, as well as Mare Island Naval Shipyard, and the Department of Labor (OSHA). In addition, he wrote and directed the preparation of significant sections of the Naval Sea Systems Health & Safety Control Manual, which has been used at all Naval Shipyards since 1980. He has also taught at California State University, the University of California extension programs, and at various professional conferences and seminars, including several international congresses. Other publications have included articles on the environmental risks associated with real estate transactions and building safer jobs.

Mr. Noble has an MBA in Health Services Management from Golden Gate University, and both a BS in Biology and an MSPH in Environmental Health from the University of South Carolina. His credentials also include being an Associate Risk Manager, Certified Industrial Hygienist and Safety Professional, and a Registered Environmental Assessor. He has served on the boards for Liberty International Risk Services, a hospital, and three environmental companies. Formerly, he was the Director of International Consulting Services for Liberty International, and he is currently an officer with Science Applications

International Corporation (SAIC), and Division Manager in their San Francisco office.

Very active in his community, Mr. Noble has participated in Peace Builders, a youth violence prevention program. While in California, he was co-chair of the Solanno County Cancer Prevention Program, and a member of the county's Health Promotion Committee. He was also instrumental in the HAS's efforts to create California's cigarette tax initiative. Mr. Noble can be reached at www.corporatewarriors.com/noble, or at noblem@saic.com.

Acknowledgments

There are many people to thank, but the one person I would like to acknowledge first is my illustrator Pauline Hauck. Pauline created drawings that gave a serious book a sense of lightness and fun. I sincerely appreciate her efforts and hope that those of you who see her work will contact her.

Pauline lives in the foothills of California with her husband and dog. She is a registered nurse by day and a cartoonist and illustrator the rest of the time. Her web site features examples of her work and she can be contacted there for additional information by going to http://www.customcartoons.com.

Over the course of my career many people have influenced me and have served as mentors. I would like to thank all of them, especially my former boss, the late Carol Tatum, who walked me through the systems safety approach used by the Navy's Nuclear Power and Radiation Control Programs. Much of my understanding of process control, quality management systems, and systems safety is the result of "lessons learned" while working for a nuclear shipyard. I am especially thankful to Tony Pattitucci and Ernie J. Scheyder who asked me to chair the shipyards Management Enhancement Team and visit all of the companies discussed in Tom Peter's book, *"In Search of Excellence."* The purpose of these visits was to learn about each company's quality improvement processes and management development programs.

The results of the research into these companies and of Deming, Juran, and Crosby's Quality Processes significantly impacted my perspective of Quality, Environmental Health and Safety management. The "lessons learned" and perspectives developed during those years has strongly influenced this book. In addition, I need to thank Bill Priess the Shipyards Production Engineer and Quality Management Czar. Bill is a great engineer and was a good mentor teaching me how to get things done given the bureaucracy and politics of a large facility like the shipyard (more than 13,000 employees), and in implementing organization-wide programs that ultimately affected more than 75,000 employees.

I also need to thank Bob Paedon, a long-time friend and nuclear quality control manager at the shipyard and Rachel Dondero who was a valuable friend and mentor for human relation's issues and some of life's lessons. Last but not least I want to thank Harry Mann who required the Occupational Health and Safety Department to oversee the shipyard's environmental program, another significant and very influential experience that ultimately lead to my becoming a co-owner of an environmental management consulting company.

In addition I would like to thank my reviewers whose long hours of work and constructive criticism have been very valuable; those people included Joe Cascio, Vic Johnson, Mike Williams, Diana Kollin, Bob Wilson, Frank Mastrioanni, Steve Levine, Kathy Seabrook, Mark Kline, Lloyd Brown, Jim Griffin, and Huw Andrews, and some of Wiley's reviewers, especially Mr. John Polhemus of Bayer Corporation.

I would especially like to thank Joe Cascio, Steve Levine, Kathy Seabrook, Vic Johnson, Bob Wilson, Frank Mastrioanni, Jim Griffin, Huw Andrews, and Mark Kline for their contributions to the Foreword of this book. Their insights and contribution have been very helpful, and will assist most readers in better understanding the practical application of an Integrated Management System.

Gratitude is also owed to Patricia McDade, the founder of the Entrepreneurial Edge, and Bill Rodgers, my management coach while I participated in "The Edge". Their insights into life and business have contributed immeasurably to the writing of this book. My long-time friend Roxanne Fynboh and my former partner Dr. Bill Cornils also participated in "The Edge"; they have both been valuable "coaches" and great friends over the years. I wish them the best with their company California Industrial Hygiene Services.

An advance debt of gratitude is also owed to John Cook, President of the Environmental Careers Organization for offering to list the book on their web site. In addition a debt of gratitude is owed to my editor Jerilynn Caliendo, who has provided immeasurable help in having the book reviewed, edited, and published. I am looking forward to working with her and Wiley in the marketing of this book, and in future endeavors.

Last but not least I want to thank my wife, Brenda, who put up with me working weekends and nights on top of an already busy schedule to write this text. Without her patience, I would have stopped long before the first draft was finished.

Introduction

Webster's defines a master as "an artist, performer, or player of consummate skill. . .a great figure of the past. . .whose work serves as a model or ideal. . .to gain a thorough understanding. . .one that conquers or masters. . .one having control of an animal." Organizational Mastery is one of conquering and controlling the multi-headed dragon that exists in many organizations. It is about being the best and ensuring that your organization understands and controls process risk; and provides its internal and external customers with exactly what they need, or require. It demands that managers strive to be "peak performers" or "players of consummate skill", models who thoroughly understand and consistently delivers the customer's expectations.

Mastery also means having a mission, a sincere gut-level focus to be the best. I have a friend, Frank Mastrioanni, who is a SrVP and Risk Service consultant at Marsh, Inc. Frank is a master not only in his career, but also as a Black Belt in karate. He is one of the elite, one of the best in this sport, and, given that he is still actively engaged in it after 30 years, you must assume that karate is one of his passions in life — an opportunity to be the best that he can be!

Organizational Mastery is also demonstrated in people like Jack Welch, General Electric's CEO, who is leading the Six Sigma commitment for GE and the Six Sigma "Black Belts", who are saving their companies hundreds of millions of dollars with a very advanced level of Statistical Process Control (SPC). We will talk more about SPC and Six Sigma in Chapter 7.

This book focuses on the process of taming the dragon, as if it were a pet, and controlling it so that it behaves in ways that are desired, expected, and predictable. The many heads like a the multi-headed dragon tend to fragment an organization, each trying to pull the system to serve its needs. Sometimes the different heads of the dragon, or of an organization do not know about or understand the needs of the other heads. This fragmented approach to process control can be changed! To get started we simply need to get out of our respective "professional boxes". Managers must learn to identify and better define the multiple root causes of process nonconformance to effectively design controls or to change procedures to mitigate process risk. The greatest level of effort must initially be applied to problem definition and gap analysis and risk assessments so that appropriate and prioritized controls and process changes can be effectively implemented. Like any other problem, or as any engineer will tell you, more than half of the "the battle" is won, once the issue is adequately defined or understood.

This book is about using the best of the various process management tools found in quality, environmental, and safety management for an Integrated Management System (IMS) approach to process risk management. QEH&S management systems can be more effectively managed if they are integrated into a single coherently managed system or process. We could discuss the advantages of various organizational structures, including matrix management. Although I believe that the matrix approach makes the most sense, how a company is currently organized will dictate the organizational structure for an IMS, therefore we will not spend much time discussing these options.

However, I believe that most organizations will benefit from establishing what I will refer to as a process enhancement team (PET). The PET is an iteration of process risk assessment groups. It includes department or work center team members, supported as necessary with other engineering or organizational participation as necessary (a Matrix approach to organization). Each department or work center will have its own PET, however, as necessary members of other departments will be required to temporarily support other work center PETs. This matrix-type support is necessary to help clarify customer (external and internal) requirements or to participate in the review/approval process for various control options, including setting schedules and determining the cost—benefit of control options. This process will require significant employee involvement. The responsibilities of a typical PET team could include the following.

- Defining and documenting internal and external customer needs
- Performing process risk assessments and reviewing existing loss or nonconformance data to start the process of identifying primary process risk factors and systems gaps
- Supporting QEH&S audit teams with their gap analysis
- Identifying and implementing prioritized process change, including the need for changes in or the scheduling of delivery for various supplies and services, tools, equipment, etc.
- Involving needed support departments and service providers in the risk assessment and processing enhancement effort
- Processing measurement and evaluation support

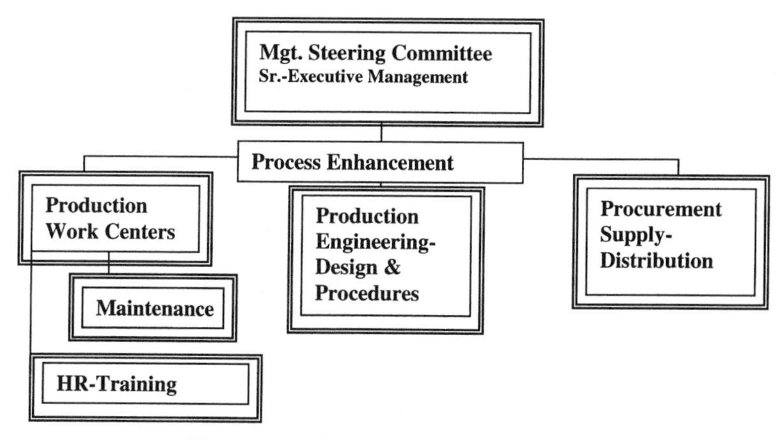

Figure 1 *PET Organizational Chart-Model.*

The PET is a grass roots structure similar to quality circles of the 1980s. They will oftentimes be responsible for performing qualitative risk assessments. Their charter is to ensure that there is full employee involvement in the assessment and risk ranking of work center processes and for recommending various corrective or process improvement actions, including behavioral safety or performance management systems. The PETs also play a very critical role in the development of critical process observation checklists and in conducting many of the routine job observations, including reinforcing desired behaviors and correcting or coaching to correct undesired performance.

Strong employee involvement is key to the success of these teams, and the overall QEH&S implementation process. If you believe that your employees are your most valuable resource, and that they are the facility experts regarding their jobs, and the barriers to peak performance and process quality, then why wouldn't you solicit their input in evaluating process risk and the development of process controls?

IMS's will be very attractive for global multi-facility companies. The IMS will standardize many procedures and will be much less expensive to implement, while ensuring greater consistency between countries and facilities implementing the integrated processes. This standard in turn will provide a higher level of comfort for corporate attorneys, risk managers, CEOs and their board of directors. Inconsistency will be the Achilles heel in any company's legal defense, especially inconsistencies in QEH&S process control and/or policy.

This book can be used as blueprint for an integrated QEH&S system. However, I believe most companies will find it more useful as a resource that identifies considerations and provides tools for enhancing or integrating existing systems. I believe that because of industry, corporate, national, or cultural reasons a specific model that fits all businesses is difficult to conceive of at best, and is certainly not the intent of this book. I would like my readers to adopt the concepts or tools we discuss and those that make the most sense to them. The bottom line is that I see this book as an argument for integrating management systems and providing tools and resources, versus a fit-all management system.

The flow chart at the end of chapter one and the suggested actions at the end of subsequent chapters are included to help you with the design of your IMS. Chapters 3–9 and the appendices provide you with technical or managerial information or some examples of the tools available to help you evaluate and manage your organizations risk. This book is not intended to provide another suggested management system standard. I have simply tried to provide concepts, tools, and hopefully the encouragement that will assist you with integrating your management systems. Tools that will be discussed and recommended for use in the IMS include the following.

- Incident or nonconformance multiple root cause analysis
- Statistical process control (SPC)
- Risk assessments
- Systems safety: management oversight risk tree analysis (MORT), fault tree analysis, change analysis, and barrier analysis
- Risk mapping
- Process flow charting, including Ishikawa, to help define desired/expected process flow, known, or potential process deviations and risk
- Risk management, including risk transfer
- Critical task observations and job–task analysis: performance management or behavioral safety
- Audits
- Emergency planning
- Contingency planning
- Evaluating control alternatives using finance tools such as present value (PV) and future value (FV) of cash flow, return on investment (ROI), payback period, and cost–benefit analysis with tools such as Liberty Mutual's savings equation, or the ORC Cost Benefit tool.
- QMS's, including Deming's, Juran's, Crosby's, and International Standards Organization (ISO) (Note that ISO 9000 and 14000 will soon be using the same basic structure. These systems also have much in common with Occupational Safety and Health Administration (OSHA) VPP, the new OSHA OH&S system standard, and a variety of standards around the world, including Australia's Work Safe, and the United Kingdom's standards for OH&S, including the draft 18001 standard, quality and environmental management systems.)

The PET concept requires active employee involvement and is critical for the successful implementation of the IMS. Without tapping the experience of this Stake Holder (your most valuable resource), the identification and ranking of risks and their associated controls will be very difficult. Each PET should be trained to understand the technical concepts of an IMS, and the resources/tools identified above.

To ensure that I do not offend any of my readers I want to define a term that will be used repeatedly in the book: *substandard conditions*. In this text we will assume that substandard conditions are procedural nonconformances, deviations from the process that management believes is executed. It does not necessarily mean that your facility is antiquated or second-class in any way. Substandard conditions does not refer to sweat shop conditions; in fact very modern facilities can have substandard conditions as defined here. Substandard conditions oftentimes reflect a breakdown in communications as expressed in poorly designed or outdated procedures and inadequate training and communications from supervisors and management. Substandard conditions and incompatible goals create the work environment supervisors and employees must work in and are the root causes leading to undesired employee behavior, accidents, customer complaints, environmental liability, and rework.

This book is about integrity in the workplace. It's about eliminating blame and creating an environment that progressively removes substandard conditions and the environments that generate quality problems, rework, environmental liability, and injuries. It's about management responsibility and accountability and aims to show managers a way to enhance their work environments and employee/customer satisfaction, while improving the bottom-line. This book is for all of us who want to enjoy our work and realize our potential and the contributions we can make to our businesses and customers. It is intended to help managers integrate QEH&S management systems and critical work processes, thus allowing more effective management in today's environment of reduced resources.

In Chapter 1 of Demings's book *The New Economics for Industry, Government, Education*, Deming states, "In quality management, the manager's functional role is tightly defined. The job of the manager is to work on the system, to improve it with the worker's help." I have always seen this situation as a need for managers to treat employees as their internal customers, and therefore vigilantly try to identify and anticipate the needs of these customers so that the manager and the organization as a whole succeed.

Management must clearly understand the work process and what is required with regards to the resources needed in the existing system, including training and communications, materials, procedures, tools, equipment, facilities, and the maintenance of all of these. How the system is designed to support the appropriate application of these resources in an integrated QEH&S management system and is one of the key objectives of this book. Deming, in discussing his 14 points for management, refers to this system as principles for transformation for Western management. Here are Demings's 14 points.

1. Create constancy of purpose towards improvement of product and service.
2. Adopt the new philosophy.
3. Cease dependence on mass inspection to achieve quality.

4. End the practice of awarding business based on price tag.
5. Improve constantly and forever the system of production and service.
6. Institute training on the job.
7. Institute leadership. (The goal of the supervisor should be to help people, with provision of procedures, tools, equipment, etc. to constantly improve the process for delivery of goods and services and to ensure that communications — especially process communications — are clear, accurate, and enhance the production process.)
8. Drive out fear.
9. Break down barriers between departments.
10. Eliminate slogans, exhortations, and targets for the work force asking for zero defects and new levels of productivity.
11. Eliminate work standards.
12. Remove barriers that rob workers, management, and engineering of their right to pride of workmanship.
13. Institute a vigorous program of education and self-improvement.
14. Put everybody to work to accomplish the transformation.

This book was not intended to present a thesis on ISO 14000, ISO 9000, or the draft OHSAS 18001. For a detailed information on these systems, obtain a copy of the standards and/or purchase one of the references to these systems such as Joe Casio's book on ISO 14000. My objective with this book is to provide the reader with a foundation for a quality management system (QMS) or an EH&S management system. I believe that this can be consistent with any of the ISO systems; TQM; the systems of Deming, Juran, or Crosby; the draft occupational health and safety management systems; or even the OSHA VPP or draft management system standards.

Many U.S. companies have chosen not to seek ISO certification but still implement a management system that is consistent with ISO requirements but not burdened with the documentation or administrative costs generated by some registrars. Therefore this book is more broadly written than most books that address specific management systems. The discussions provided here are intended to clearly illustrate how all of these systems have a common management and process engineering foundation. The key for success is to implement an IMS or to create an IMS by merging the missing pieces of QEH&S Management.

This book encourages managers to implement quality or environmental management systems and, to the extent possible, integrate these systems, including process safety. References to the various quality and environmental management systems, including ISO 9000 and ISO 14000, are simply to give the reader a framework to build an effective system that is consistent with processes used globally by their customers and vendors. Becoming ISO certified is a business decision that will not be discussed in this book. My overriding objective is to help you improve process quality while reducing EH&S risks.

However, I will encourage you to consider using these systems to establish strong process quality management. I also believe that if your primary focus is on improving performance, and not on certification and regulatory compliance, then the burden of documentation and regulatory conformance will be significantly reduced. Your system should first of all make sense from an operations and process improvement perspective and not be driven by concerns over your audit. In addition organizations can work with their information system (IS) departments to develop the record keeping and data management systems, policy, and procedure templates that will need to be shared between various elements of the organization. Some data and information management systems that are commercially available can significantly reduce the documentation burden associated with ISO certification, and they should be evaluated in light of the organization's business needs and existing systems.

A well-thought-out QMS that integrates EH&S will significantly improve process performance and reduce risk and the costly liabilities associated with process nonconformance over time, even if certification is not a priority. The decision to implement a certified management system should be a second priority and not the focus of the implementation effort. Key to improved performance is system design and tapping the resource that exists with the first-line supervisors and employees who perform the organization's daily critical tasks. If employees are your most valuable resource then they must be heavily involved in the QEH&S process improvement system.

This book should help operations managers, quality engineers, risk managers, safety managers, industrial hygienists, and environmental engineers see the power of working together to improve process performance at all levels. The tools and concepts recommended here can become part of an overall decision-making strategy to help manage and mitigate risks, including some discussion of sharing risk or risk transfer.

Most of the tools discussed or provided here are not new. As an example, SPC has a 50-year history. Systems safety goes back to the 60s and 70s with the Department of Energy. Risk assessments and facilitated group processes for evaluating practices has also been used for various management improvement efforts, including safety and quality.

Performance management or behavioral safety is newer, but it has gained a lot of credibility and has been implemented in a wide range of business activities over the past 15 years. What is new or unique about this book is compilation of the various tools and systems discussed above to ensure that there is a more coherent and cost-effective approach to managing QEH&S and EH&S.

Deming and Juran drew their experiences from World War II. The systems used by manufacturers supporting the war effort focused on statistical control of mass production to supply the military. Lives were at stake, the fate of the world was at stake, and quality-on time delivery of goods and services was critical. Deming and Juran then used these methods to help them with the post-war reconstruction of Japan. For some reason, American manufacturers believed that the post-war production of products did not require the same rigor as the military-industrial output. This notion was probably attributed to some arrogance over having the only significant industrial infrastructure remaining in the world, and to a return of men to production jobs who were not familiar with these systems.

The boom years of post-war America focused on quantity, and taking advantage of increasing market demand. The Japanese, however, were starting over and had focused on developing products and the systems to deliver them, which balanced both quality and quantity.

Process management is becoming very important for companies not only in the United States, but both in Europe and in countries with developing economies. I am asking the reader to think about integrating QEH&S programs in their facilities. I believe this objective is possible by using ISO 9000/14000 as potential models to build the integrated system or management process.

Chapter 2 reinforces the arguments for IMS's and focuses on safety management, which until recently has not received the attention it deserves in the ISO management standard setting process. Currently there is a draft occupational health and safety management system standard OHSAS 18001, that most likely will become part of a future ISO standard, and even an integrated management system occupational health and safety standards, such as BS 8800 standard (IMS). It is interesting to note that in the United Kingdom, management systems have been required by standards such as BS 8800. Introduction to measurement and behavioral observations for quality and safety will be introduced, in several places in the book and discussed in some detail in Chapter 7. The discussion in this book is simply to show how these tools can powerfully improve job observations and performance measurement.

I repeatedly emphasize the roles of senior managers and supervisors and the need to establish policies, goals, and objectives, including frequent follow-up as part of the total process management process. Planning will be discussed as the foundation for designing and successfully implementing a total process management system. Clear roles and responsibilities, as well as a process for systematically reviewing process/resource inputs to the system such as the procurement procedures, training, maintenance, and procedural controls, must be established.

The concepts of multiple root cause analysis, risk management, risk mapping, process nonconformance and accident costs, and systems safety are introduced and discussed in several sections of this book. These concepts serve as tools to help prevent nonconformance and to learn from existing nonconformances. Processes have primary or critical process inputs (see Fig. 1) that, if substandard in any way, they will ultimately be the root cause of nonconformances and incidents.

It is our responsibility to evaluate all primary process inputs to ensure that they are adequate and that they do not become the root cause of an incident. We are also responsible for learning from our mistakes and evaluating process nonconformances and incidents to identify the multiple root causes of the process failure. These responsibilities must include taking appropriate actions to ensure that failure does not happen again, including sharing lessons learned with appropriate managers and employees so that they to can avoid recreating the incident.

Measurement and evaluation is discussed in Chapter 7 as critical elements of IMS. Audits, including initial gap-analysis assessments, managerial, technical, and regulatory, are discussed. On-going systems evaluation and performance improvement-reinforcement is encouraged by using tools such as performance management or behavioral safety or other critical task-process observations tools such as job-task analysis.

The necessary elements and implementation of an integrated total process management system include the following.

- Employee involvement
- Training
- Communications
- System documentation and document control
- Records
- Design control
- Purchasing
- Hazardous materials/waste management
- Management of contractors and vendors
- Emergency preparedness and contingency planning processes
- Use and maintenance of equipment, tools, and the facility

Operational control as discussed in this book demands emphasis on the use of procedures to control the work at the process resource input and design stages. Critical to understanding process improvement and incident prevention is understanding the management system and the elements of the system that are causing process nonconformance and incidents.

The concept of multiple root cause analysis and the tools for process nonconformance or incident investigations, including the need to document the total costs of nonconformance (direct and indirect costs), will be discussed in detail in Chapter 8. Until management understands the financial impact of substandard process design/execution and the benefits of process change communicated in financial terms such as "return on investment", "payback period", very little will be accomplished. Otherwise, "bandages" will be applied to the process enhancement or corrective/action effort.

Identifying the causes of process nonconformance and sharing lessons learned to correct/prevent a reoccurrence is one of the most important tools supervisors and managers will have on the path to continual improvement. If you know what went wrong and why without blaming employees you can improve the process and prevent a reoccurrence.

I also believe that audits, and the follow-up of corrective actions, will be critical to successfully implementing any management system. The appendices of this book contain various management and technical audits that can be edited or merged as appropriate to start the development of your organization's audit tool. These tools are presented only as examples. I would encourage you to edit or merge them to more accurately reflect the needs of your existing or proposed management system. This recommendation is especially true of the check sheets provided in Appendix II, which only point out key considerations, without specific regulatory, facility, or corporate policy considerations.

We need to get away from stand-alone safety or environmental programs and policies and instead need to ask ourselves the following questions.

- What can we do to improve the process, reduce rework, breakage, and worker's compensation claims while increasing productivity; efficiency; profitability; and customer, management, and employee satisfaction?
- How can we involve employees and management in creating a process that saves money and increases profitability while reducing waste, rework, and losses due to accidents, as well as process nonconformance to requirements?

Because of my 25 years of experience as an EH&S manager, I have started this book with a discussion focusing on EH&S. However, this focus on the work process is only for illustration of the inefficiency created by the fragmentation of process management. I am also trying to create a challenge for managers who are struggling with quality and the increasing costs of injuries or environmental problems. What can you do to shift the existing process paradigms? How can the work process be simplified? How can we improve the process, reduce redundancy, and identify and eliminate QEH&S risks? Are you willing to invest in defining critical process performance observations check sheets and job performance reinforcement, and corrective actions?

Two of the primary keys to improved performance are the system design and frequent job observations and reinforcement of expected or superior performance, and positive correction of substandard performance. Correction of substandard conditions with even interim controls must be a high priority. In addition we need to understand the positive or negative reinforcements and consequences that cause employees to deviate from management and process design expectations? Understanding these consequences can be instrumental in understanding some aspects of process design risk and the need for better product or production process design to QEH&S performance.

I worked at a shipyard with more than 13,000 employees for nine years and observed that the key to the Navy nuclear programs' success was its managers' understanding of process management and effective use of audits. I think that most people who know anything about nuclear power will agree that the U.S. Navy's program under Admiral Rickover was and remains the best in the world. Many people however probably do not understand the level of personal control, through a network of auditors, and systems safety, that was exercised by Admiral Rickover.

This book consists of ideas and approaches that can work with separate systems or IMS's. However, I believe it is important to eliminate the fragmentation of process control caused by separate environmental or safety programs. It oftentimes appears that disjointed process management systems contribute to supervisor overload and system failures.

Not long ago I was in China, discussing the Three Gorges Project with some associates and asked what it would mean to the Chinese people. At one point they described Shanghai as the head of the dragon with its heart in the Yangzhe at the Three Gorges Dam project, and its body or tail stretching inland along the banks of the river. At some point this metaphor was born again as a way for me to describe businesses and operations management.

I see the head of the dragon as senior management, who must be committed to the process and must establish policy and provide the resources necessary to successfully implement a process of continuous improvement. The heart of the dragon represents the first-line managers and supervisors, especially those who have production or business development responsibility. However, many organizations have a dragon with multiple heads (various departments or systems managers), each working against itself. This disjunction in most instances is not malicious, but it may be a manifestation of management competition. Controlling the dragon and organizational mastery require the various heads to work together to identify and control risk, while creating value in each process.

Without a well-thought-out plan integrated across organizational lines with authority from the head of the dragon, process nonconformance and incidents will occur. The integrated plan or process must then be effectively translated into specific tasks by its heart, or supervisors. Without this factor, the body of the dragon tends to be neurologically or physically disconnected.

In organizations with this type of disconnection, the dragon's tail, or the organization, tends to thrash about uncontrolled in the marketplace, causing injury to itself, its customers, and other stakeholders. The goal of this book is to help management lead through its heart to control the dragon and to significantly increase the involvement of its real strength in the employees of the organization. To be truly successful you must also maximize the effectiveness of well-trained supervisors by providing them with the resources needed to perform their jobs safely, on time, and within budget.

Without good supervision the dragon has no value and no business. Inferior or late delivery of products and services, will also ultimately lead to its death. Like any heart it must be exercised, nurtured, and developed to ensure that the dragon remains healthy. The tail of the dragon, the people and the process used to produce products and services, includes all of the resources that are applied to the process employees/contractors, facilities, tools, and equipment.

Another problem in organizations is their tendency to focus on regulatory compliance versus process management. Some organizations have implemented a management system only to become frustrated with it or question its value as they focus on documentation versus process management. In other situations, they become lost in quantitative risk assessment data and can no longer see the forest through the trees. Simple common sense heads south for the winter!

1. **Throw money at it, sometimes this cures the problem.**
2. **Kill it, and start all over, certainly not an option most managers would choose!**
3. **Control the beast, make him work for you, take advantage of its power!!**

Figure 2 *There are three ways to control an organizational dragon.*

This assessment does not mean that you should ignore regulations or documentation. You must conform to both, however, regulatory and documentation conformance should be a byproduct of good process management and not the primary focus of the management system.

Over the past 15–20 years mangers have been bombarded with a variety of management programs and processes, including ISO 9000 Quality Management Systems and ISO 14000 environmental management processes. In addition there has been a variety of safety systems (some legislated) and an assortment of quality or process management programs.

If you look to the European Union, and the United Kingdom in particular, you see regulatory bodies adopting the ISO framework for EH&S. This trend can be seen in other countries around the world such as New Zealand and Australia, and it clearly reflects a different approach to the prescriptive regulatory one historically taken by the U.S. Environmental Protection Agency (EPA) and OSHA. An example of a management system approach can also be seen in the draft OHSAS 18001 standard (Ref. 22).

Last year, the American Society for Quality's Audit Committee reported in its April newsletter that U.S. delegates of the Quality Technical Committee TC 176 and the Environmental Committee TC 207 have expressed interest in combining the management systems and respective audits.

A management system is intended to help facilities establish processes, procedures, and metrics that allow them to perform tasks right the first time, within budget, on time, and with no rework, redundant procedures, injuries, or environmental liability. Companies may need technical experts to help them with specific technical issues around QEH&S, however integration of these systems is quite feasible.

IMS's will be critical to the success and competitive posture of companies in the future. Management must align its resources (people, tools/equipment, facilities and maintenance, procedures, materials, vendors/contractors, training, and supervision) to optimize QEH&S performance.

Oftentimes, programs or processes have their roots in the works of Deming (Deming, 1982) and Juran, or in systems safety, which originated with the Department of Energy (DOE) and National Aeronautics and Space Administration (NASA). For those unfamiliar with systems safety, this process is not focused simply on worker safety. Systems safety includes an assortment of risk assessment tools that help management identify and, to the extent possible, quantify potential process risks and to ensure operational readiness. This process could include evaluating business risks as well as the risk of environmental contamination or injury to employees or customers.

Another valuable tool for communicating risk to management is "Risk Mapping", developed by Microsoft. "Risk Mapping" is a way to summarize the frequency and severity of various risks, including the level of risk transfer and the impact of proposed controls for defined risks.

The key to all of these systems is enabling managers to tap the expertise of their internal customers (their employees) and to clearly define their existing work processes and associated process risk. It also means that management with employee involvement must identify tasks or steps in a process that by its execution has tended to make work less efficient or difficult to do safely. Management must also understand what influences employee behavior.

Many programs are designed to reward or recognize those who don't have injuries. More permanent programs focus on what people do (or don't do) that cause injuries, compromise quality, require rework etc. A total of 85% of safety and process problems can be resolved if the processes are well-designed and if routine job observations provide immediate reinforcement and feedback of desired behaviors, and immediate corrective action for nonconformance. The supervisor and trained employees are in an excellent position to coach and reinforce the desired behaviors to ensure total quality and EH&S control.

Generally nonconformance with requirements consists of behaviors influenced by such things as production quotas that don't allow procedural compliance, use of poorly designed tools or equipment, or tools that have been modified. If the system is designed properly and is operated within design specifications, then employee compliance with expected behaviors or standard operating procedures will be greatly enhanced.

If the management system is weak, numerous incidents of process nonconformance will occur, which will ultimately result in a serious accident, or process nonconformance, given the right combination of events and root causes. Keep in mind that root causes refer back to those primary process inputs such as communications, design, tools, materials, and so on, as illustrated in Figure 1. Inherent in this statement is that changes to a process, including increases in production rates, should be evaluated to identify risks that might have been created by the change.

Unfortunately we have too often created confusion in the workplace by not consistently sticking to a chosen plan for process or quality improvement. We have jumped from Deming to Crosby's cost of quality(11), TQM, and/or reengineering. Some versions of these quality programs have been adopted internationally in ISO quality and environmental standards. As mentioned earlier, some countries have been adopting versions of ISO as a management system standard for EH&S.

The reason ISO has become so popular globally is that, just like common thread sizes on bolts, managers of multi-site or global operations have recognized that having multiple systems is dysfunctional. Also, the corporations may create additional liability by not consistently applying corporate policy. The biggest problem for managers is that, like specifying thread sizes on bolts to ensure consistency and sustained delivery of these products, common management measurements and language are needed. Too often we tend to adopt metrics and language to describe various programs that are not consistent, or we tend to fragment process management into artificial compartments such as quality, environmental, or occupational health and safety. Another reason management systems have become popular is because they are seen as a way to control the escalating costs of process nonconformance. Six Sigma, the newest and most rigorous of systems, has a goal of defect elimination, or no more than 3.4 defects per 1,000,000 opportunities.

Dramatic increases in medical and workman compensation costs, abuses, and, in some cases, outright frauds have caused employers, medical providers, and insurance carriers to look at what can be done to reduce costs. In 1997 occupational injuries alone cost $127.7 billion in lost wages and productivity, administrative expenses, health care, and other costs, not including the cost of occupational illness. (National Safety Council report, "Accident Facts"). Medical costs alone were $20.7 billion, and workers' compensations costs were $43.5 billion. Each worker in the United States must produce and average of $1,000 worth of goods and services to offset the cost of work- related injuries and illness. Every 5 seconds a worker is injured, and every 10 seconds a worker is temporarily or permanently disabled.

All accidents nationwide cost ~3% of the gross national product (GNP). When this loss is expressed in medical expenses, lost wages, insurance claims, production delays, and equipment downtime, it significantly reduces business productivity and profitability. A more costly problem that would often be related to the same management control issues is rework, which would easily more than double the injury losses. Similarly the United Kingdom estimates that accidents wipe out 5–10% of industrial trading profits annually (View Point, Corporate Cash Flow, April 1995).

Potentially more costly problems, which can often be related to the same management control issues, are quality or rework, and environmental liability that can again easily more than double the losses due to workplace injuries. Indirect costs associated with incidents range from 4 to 10 times the direct costs of wage loss and medical cost, greatly increasing the significance of this type of process failure.

The problem oftentimes is attributed to a lack of understanding of process management, coupled with our tendency to compartmentalize process quality, safety, and environmental management. Who benefits by this fragmentation?—the line supervisor, production management, or the specialists who manage these departments? Too often these program managers consider these processes to be the center of their corporate universe, and their peers' goals to be less important, failing to take advantage of the opportunity to work together.

We need to back up and look at work processes from our internal customer's perspective, as well as the line supervisor's and work center employee's. As a senior manager or manager of a quality system or EH&S program ask yourself the following questions. What is the line manager or supervisor's primary mandate or concern? What is senior management holding the supervisor accountable for? At the end of the day is management accountable not only to its external customers and shareholders but also to its internal customers, supervisors, and community stakeholders?

It is management's job to understand the corporate objectives and the needs of its customers. It is also the responsibility of management to ensure that it remove all roadblocks from the paths of supervisors and employees. Continuous improvement should mean having a constant focus on making each task in the production process as simple and efficient as possible, by eliminating redundancy, and correcting facility process designs whenever they are identified.

True understanding and appreciation of their internal and external customers and the summation of incremental improvements over time are what sustain companies and their profit margins. "Sustained focus on these values and objectives" must be the management mantra for the new millennium. Many companies lack focus and have conflicting objectives, as is evident in redundant and multiple management systems for QEH&S. The other problem contributing to lack of focus is when leaders delegate responsibility and are not actively involved in the process enhancement system. Too often it is believed that the EH&S manager or the quality manager can cover the same territory as hundreds of supervisors.

Why is it that QEH&S programs oftentimes appear to be the responsibility of the quality or EH&S department or a distant corporate entity versus the production manager who controls the work process? Why is it that workplace process control instruction or designs oftentimes fail to include QEH&S? Why aren't quality managers and EH&S working more closely together?

Quality and EH&S managers must also be able to document both the direct and indirect costs of process nonconformance. This step is essential if the risk or process manager is going to be in a position to favorably influence management regarding system or work process improvements or loss-control efforts. In addition, the process managers must be prepared to present capital or system expenditure requests in terms that the operations and finance managers understand. A 1994 study by Liberty Risk Services in the United Kingdom found that 70% of the corporate decision-makers for health and safety had the title of finance director or above. Yet most of these managers do not have accurate data on the costs of accidents.

Does the concept of continuous improvement apply only to production rework? Why wouldn't you consider environmental damage or employee injuries a process failure? Quality failures often have safety and environmental implications. After all aren't our employees our most important cutting-edge resource, or do we see them as disposable and easily replaced? Through painful experience and financial consequences many companies have learned that environmental quality control is necessary to eliminate potential liabilities that can cause the financial failure of a company, not to mention potential criminal liability. In fairness to operations managers, too many safety or EH&S managers have not understood how to address process issues effectively, especially in financial terms.

These questions may seem sarcastic or even negative, but they are intended to highlight the dangerous attitudes exhibited by some managers and supervisors who don't understand process control. If any company in today's global market expects to be competitive it must be willing to consider integrating its quality and EH&S management systems. EH&S managers must also learn more about quality management and communicating to production and financial managers in terms that they understand. Changing metrics and reporting losses or risks in financial terms, or considering returns on investment when recommending process controls or process changes, is critical if we expect to bridge the gap between production, quality, and EH&S management.

More to note is why employees or supervisors who control these work processes are often the last ones to be contacted for review of work process designs or process controls? Employee and supervisor involvement in process enhancement risk assessments and critical job observations are essential if we expect to mitigate QEH&S risk. After all, they are the ones who live with the work process, good or bad, and too often struggle with the results of management's failure to consult with them. One thing that all managers must be aware of is that in implementing any organizational change, even positive change will cause some system-personnel stress. Therefore, change management of the cultural and process changes must be an integral component of the planning and implementing process.

Realistically, how many of you or how many of your employees go to work with the idea of doing bad work, sabotaging the work process? Perhaps you decide today is the day I will create an environmental disaster for my company and community, or maybe I will mutilate myself, or make the supreme sacrifice in the name of the carelessness god, by killing myself at work?

What are we trying to do with an integrated QEH&S Management System?

Managers of most mature companies understand the need for a quality management process. Industrial managers and developers by now know the risk associated with mismanagement of hazardous materials and waste. The difficulty has been how do you do it, which process TQM, Deming, Juran, or Crosby. etc.

In many ways environmental issues appear to be the most difficult because of the technical complexity associated with re-mediation projects or the design of new chemical processes. Then there is occupational health and safety that many managers think is out of their control. They believe many injury claims are fraudulent but difficult to prove, or that the employee was careless. If they only had more conscientious employees the safety problem would go away! In reality 80–85% of the process failures, including injuries, are caused by process common cause, variations built into the process by the existing management system.

The typical company spends one week per employee to implement a TQM program. The biggest mistake is a failure to focus on a critical few issues, and too much emphasis is focused on documentation and regulatory compliance versus process and task management. The average PET has five people and their compensation averages $50/hr. They will also utilize 250–750 hours to solve problems at a cost of $7,500–$22,500. The PET must be focused and working on prioritized projects, and the critical elements of work processes causing nonconformances. Parietos Principal appears to apply here.

One of the objectives of this book is too show managers that all of these processes are under their control, and that there is tremendous benefit to integrating the management process for controlling these risks. Using ISO 14001/9001, I hope to convince you that a common process management system QEH&S, integrated and built on ISO can be very powerful. The ISO standards and their various derivatives provide a framework consisting of a number of interconnected process management tools and services organized to integrate quality and EH&S management.

The ultimate goal is to reduce time to perform tasks and to create added efficiency while reducing costs, waste, and personnel injuries. A key tool in any QMS is its qualitative risk assessment system, and risk mapping process, both of which are discussed in more detail in Chapter 5. Quantitative risk assessments and detailed fault tree or systems analysis are needed, but in many instances they should be used only after passing through the sieve of the qualitative risk assessment. Qualitative risk assessments involve employees and other stakeholders and in most instances allows management to quickly identify, assess, and establish a plan to eliminate or control process risks.

William B. Smith, Vice President for Quality Assurance for Motorola, pointed out the value of employee involvement (e.g., quality circles, and process enhancement teams or in processes such as the qualitative risk assessments). It should also be noted that, like the statistical quality control process used widely by the military-industrial complex during World War II, quality circles were also widely used; but like the quality improvement process it did not survive post-war industrialization in the United States but was reinvented in Japan. In Smith's speech to the National Private Truck Council, he stated:

> In the Motorola experience it was found that 91% of problems were hidden from general management. The general manager was aware of only 4% of problems on the production floor. General supervision was only a little better off, being aware of 9% of the problems. The largest gap in the information flow appeared between the production supervisors and general supervision. Line supervision was credited with being aware of 74% of the problems. Not surprising, the workers were found to be aware of 100% of the problems.

Smith's statement is a compelling reason to ensure that your most valuable resource, your employees, are heavily involved in the QEH&S process.

1

Controlling the Dragon —The Integration of Management Systems

"A Process is a process is a process," control of the process by any other name Quality, Environmental, or Safety, is still Process Management!

Dissatisfaction with the final product or service of an organization is called trouble with quality. However, it is only a symptom of what is happening inside the firm.
—*Philip B. Crosby*, Quality Without Tears, The Art of Hassel-Free Management

Waste is worse than loss. The time is coming when every person who lays claim to ability will keep the question of waste before him constantly. The scope of thrift is limitless.

—*Thomas A. Edison*

ISO 14000/9000 HARMONIZED QUALITY AND ENVIRONMENTAL HEALTH AND SAFETY (QEH&S) MANAGEMENT SYSTEMS

There is really nothing terribly significant about ISO Management System Standards, nothing new in them, except the fact that there is a consensus on the elements contained in the standard. The value of the ISO standards is that they give companies a common foundation to discuss, and evaluate the minimum expectations in a competently designed and executed system.

The common elements establish the benchmarks for expected quality and EH&S performance. It is for these reasons that I am an advocate for including occupational health and safety in the systems, and to the extent possible integrating them. This in no way means that you will not have a need for specific environmental, quality, or safety requirements, or technical resources because you will. What it does mean is that these resources will be used more effectively. Your focus will be on prioritizing "Enterprise Risk" and in controlling the systemic causes of process nonconformance which will in most cases eliminate or reduce quality, environmental liability and safety risks by correcting the multiple root causes of process failures. All management systems including Deming, Juran, and Crosby have common principals that are identified below, and which are discussed and compared in more detail in this chapter.

Table 1 is provided to show how the quality improvement principals defined by Deming are consistent with ISO Standards and good QEH&S management principals.

Principle 1: Commitment and Policy
Principle 2: Planning
Principle 3: Implementation (operational and risk control)
Principle 4: Measurement and Evaluation
Principle 5: Management review and Improvement

These principals are specifically discussed in the United Kingdom's standard BS 8800, Australian/New Zealand Occupational Health and Safety Management System Draft Standard, (1996). They are also covered in ISO 14001 and found in a different context in ISO 9001, TQM, and the multitude of quality and EH&S management system standards emerging around the world. They are the foundation of any good management system.

Principle 1. *Commitment and policy.* An organization should define its process control policy to ensure commitment to the management system. Execution of management's policy should be evident in internal documents and procedures, including management, supervisors, and staff training. The policies should include conformance to regulatory requirements, systematic risk assessments and process control evaluations.

Principle 2. *Planning.* An organization should plan to fulfill its process control policy, goals, and objectives. This plan should include review of facility, tool, and equipment process readiness or applicability. It should also include the review or development of well-defined process control instruction for hazardous operations. Process management requires evaluation of assigned supervisors and staff to ensure competency to perform work or that they have available and use of correct procedures, tools, etc. Roles and responsibilities must also be clearly defined to ensure implementation and adherence to policy and standard operating procedures (SOPs). Inherent in this stage is the need to perform risk assessments that enable prioritizing and focusing the planning process.

Principle 3. *Implementation and Operational or Risk Control.* For effective implementation, an organization should develop the capabilities and support mechanisms necessary to achieve its OHS policy, objectives, and targets. Policy should be posted and clearly communicated to all managers, supervisors, and employees. Action plans and target dates or milestones must be established and monitored to ensure plan execution. Controls start with systems review and risk

TABLE 1. Quality Management-TQM (Key Deming-Crosby Steps) Versus Safety Program Management

Quality	OH&S and Environmental
Management commitment and organizational constancy of purpose	Same
Adopt New Management Philosophy, including Crosby's Four Absolutes: • Quality is conformance to requirements defined by the customer (internal and external, and defined in process requirements, training, etc.) • The system of quality is prevention • The performance standard is zero defects, 100% conformance to defined standard or process requirements • The measurement of quality is the price of nonconformance (direct and indirect cost) and frequency rates	Same
Must have a management system for continuous improvement	Same
Constant improvement of the system for product or service delivery	
Must not depend on inspections alone to achieve quality	
Institute formal training and on-the-job training (OJT), coaching, mentoring, and self-improvement for process improvement, which includes quality, safety, and environmental requirements	Same
Leadership: Create an open, trusting environment	Same
Drive out fear	Same
Break down barriers between departments and internal customers and understand the concept of the customer	Same
Understand and implement process for error cause removal	Same
Identify the multiple root causes of process failure	
Employee involvement: quality action or improvement teams	Safety Committee or Quality Improvement Team
Other Deming requirements include • Ending the practice of awarding business on the basis of price alone • Eliminating slogans • Eliminating management by objective • Removing barriers that rob hourly workers their right to pride of workmanship • Removing barriers that rob people in management their right to pride of workmanship • Putting everybody in the company to work to accomplish the transformation	

assessments at the design and planning stages and include all resources applied to the process: people, purchasing, supervision/management, tools, equipment, facilities, raw materials/parts or components, maintenance, and communications, including training, and procedures. These key categories of process resource input provide the tools and create the process or work environment for producing goods and services. They are also the source of the multiple root causes of process failure creating nonconformance and forcing employees to adopt substandard behaviors that ultimately lead to on-the-job injuries, process failures, and rework.

Increasingly, operational management must also evaluate the capability of contractors or sub-contractors to deliver needed products or services. Also, managers will need to develop emergency response and contingency plans to eliminate or reduce the potential for losses.

Principle 4. ***Measurement and evaluation.*** An organization should measure, monitor, and evaluate its process performance and take preventative and corrective action. This approach must include the identifying of multiple root causes of nonconformance accidents or process failure, determining appropriate process control, corrective actions, and sharing the lessons learned with other parts of the facility or organization. This practice could include behavioral job observations, audits, inspections, and safety and industrial hygiene surveys or systems reviews, including job–task analysis or systems safety reviews. Behavioral performance management or safety observations can be a critical element of this process, providing proactive measurements of systems performance versus reactive measurements such as those from accident reports or employee–customer complaints.

Benchmarking, measurement, and evaluation are essential for program success. Benchmarking compares performance against target industries, companies, and processes internal or external to the company. It is the standard against which managers measure their operations performance.

Benchmarking is always important, and there are various ways to approach this aspect. One way is to define your own meaningful benchmark, for example, benchmark against your self, other facilities, or your industry. However, in establishing benchmarks, it is important that the costs of nonconformance or accidents be much more accurately documented, including direct and indirect costs associated with losses or rework. Indirect cost must be accounted for, as they will almost certainly exceed direct cost and are oftentimes not recognized (see Fig. 2).

Principle 5. ***Management review and improvement.*** An organization should regularly review and improve its process management system, with the objective of continual process improvement, progressive risk reduction, and complying with relevant standards or guidelines. If the system is not being fully executed or is not generating expected results, then additional management attention and changes in process direction or design will be necessary to improve the system, including the establishment of new or revised goals and objectives.

Management System

The management system is best viewed as an organizing framework that should be continually monitored and periodically reviewed. It provides effective direction for an organization's process management activities in response to changing internal and external factors. A management system should not be confused with the audit tool. The audit simply evaluates conformances to best practices, policy, regulatory or system requirements. It is the X-ray to define problems, used to direct the prescriptive design improvement process. It is the execution of systems safety or process controls, including engineering controls, training, tools, and equipment, that comprise the management system. An environment of accountability and non-blaming acceptance of process management must exist. Essential elements of quality or EH&S process management include:

- Inventories of critical processes, equipment, and tools (ranked)
- Risk assessments
- Observations/analysis
- Training
- Surveys/inspections
- Audits
- Process evaluations
- Incident investigations or process failure analysis (multiple root cause analysis)
- Corrective action design of process improvement and aggressive follow-up during and after implementation
- Budget and charge-back-to-work center for process failure and accident losses/costs
- Cost benefit analysis and system for the measurement of the cost of quality

In the 1980s Phillip Crosby was popular as a quality management expert with the publication of his book *Quality is Free*. I was director of environmental health & safety at a large public shipyard during this period, and our executive management committee made the decision to begin our quality improvement journey by using Crosby's program. Crosby promoted two concepts that remain with me throughout my professional career:

Quality is conformance to requirements [defined by the customer (internal and external) and in terms of process requirements, training, etc.]

- The system of quality is prevention
- The performance standard is zero defects, 100% conformance to defined requirements

The measurement of quality is the price of nonconformance (direct and indirect cost), the cost of process nonconformance rework, and the cost to ensure quality. The cost of quality is the additional cost in the system attributable to not doing it right the first time. The earlier nonconformance is identified, the less costly it will be. When nonconformance reaches the customer, the direct and indirect costs can be staggering. The cost of prevention is low compared with the cost of error identification and correction.

Like the prioritized risk assessment process, the key to success is to first define the process and to then identify the risk of nonconformance potential in the various steps of the process. When the process nonconformance risks are clearly identified and understood they can be ranked for allocating time and resources to corrective and preventive actions. The key to added value is to significantly reduce waste in the design, production, and delivery of services or products.

About 85% percent of the root causes of waste are due to process redundancy, or rework controlled by management (common causes). Only, 15% of nonconformance is caused by indifference or employee-generated mistakes (special causes).

Cost of Quality, Average Rework (Ref. 11)

- 20–40% for Manufacturing
- 50–75% for Service Industries

On average, companies experience two days a week devoted to doing

- Right things wrong—the correct service or report is sent to the wrong or incomplete distribution or a product is not installed correctly.
- Wrong things wrong—the bill is not prepared on time and has errors.
- Wrong things right—a well-written report or well-designed product is not what the customer wanted.
- Process or administrative redundancy and rework.

What if you could convert these two days per week to solid first-time quality production? Think of how even a relatively small improvement would increase profitability. Keep in mind these savings or the results of loss prevention to directly to the bottom line.

For every customer who complains, there are 26 others who have remained silent, and the average wronged customer will tell 8–16 others. Among a population of unhappy customers, 91% will never purchase from you again, and it cost five times as much to attract new customers as it does to retain existing business. Therefore, for every complaint about 250 others will hear about it. Only 9% of customers quit over price, while 68% quit because of indifference from the owner, manager, or employees. Work process nonconformance is like the iceberg in Figure 2; many of your more costly losses are not readily apparent.

What does the term, *best practices* mean? Unfortunately, to many this term means the most expensive option, service, or product simply because it is more expensive. However, common sense tells you that this cannot possibly be true as a rule. Oftentimes you would not want or need the most expensive option. The best practice would be the most cost-effective process improvement or barrier to prevent process nonconformance and accidents and to enhance production quality and efficiency.

Visible costs of process nonconformance oftentimes are very little compared with the indirect costs such as:

- Rush delivery
- Idle time
- Duplication of effort
- Process or administrative redundancy or rework
- Unspoken customer dissatisfaction, thus the need for follow-up/validation, and verification or customers' specifications
- Low morale, unwanted turnover
- Turf battles and communication breakdowns
- Unnecessary reports, redundant and inefficient data collection
- Excess or unused inventory
- Customer complaints and lost business
- Decreased or under utilized capacity
- Warranty corrections
- Lawsuits
- Damaged public relations and corresponding loss of market share

Supervisors and operations management tend to focus on budget and schedule conformance first and maybe, depending on senior management commitment and interest, give secondary attention to quality and EH&S process initiatives. Many companies still do not adequately measure performance or benchmark in these areas.

Benchmarking is always important, and there are various ways to approach this. One way is to define your own meaningful benchmark, for example, benchmark against yourself, other facilities, or your industry. However, in establishing benchmarks it is important that the costs of nonconformance or incidents be much more accurately documented, including direct and indirect costs associated with losses or rework. Indirect cost must be accounted for, which will almost certainly exceed direct cost, and oftentimes is not recognized, see Figure 2.

Compare your company's performance to Lincoln Electric, which has been one of the most productive companies in the world for 50 years and whose employees have taken home annual bonus checks that are equal to 90–110% of their salaries for more than 40 years. Managers and employees at Lincoln Electric take quality and EH&S very seriously. In fact, they return to the plant to correct nonconformance on their own time. The culture at Lincoln Electric is that they were paid to do their jobs right the first time and, just like any other consumer of services, the company should not be required to pay twice for a requested job. Lincoln Electric has rated quality as one of its primary criteria since the 1930s.

Because measurement and accountability for quality and EH&S are in some cases not clear, the supervisor may focus on those process areas where measurement and accountability are clear and keenly watched by their managers. These factor are most often their budget and schedule. Unless quality and EH&S are given the same level of attention and have the same level of unambiguous measurements, the quality and EH&S process improvement effort will fail.

> The excellent companies really are close to their customers. That's it. Other companies talk about it; the excellent companies do it.
>
> —*Peters and Waterman, In Search of Excellence.*

Elements of ISO 9000-14000 that are consistent with Demings's 14 Principals

- Management Program: responsibility and control (management commitment, including resources)
- Quality system, policy, objectives
- Documentation and document control
- Operations control, including
- Design control and process review
- Purchasing
- Maintenance
- Material handling, storage, and packaging, including waste minimization
- Emergency preparedness
- Monitoring
- Nonconformance evaluation and corrective actions (includes rigorous follow-up)
- Audits (monitoring/measurement)

ISO 9001/14001 HARMONIZED MANAGEMENT SYSTEMS

ISO harmonized systems provide for continuous improvement and are sufficiently flexible.

- They are applicable to companies of different sizes.
- They are appropriate for the hazards and risks specific to a given organization.
- They are compatible with company or local culture, including integration (harmonization) of existing environmental or QMS's.

The ISO was established in 1947 to create international standards for the manufacturing of products that would be used by customers in different countries. International managers realized that the cost and duplication of effort by not having standardized methods could be significant. In the late 1980s ISO decided that this same principle applied equally to the management of quality and environmental systems.

Effective EH&S or QMS's have four stakeholders with their respective needs or corporate goals for improvement. These typically include:

Shareholder: Profitable growth
Society: Benefit—value, positive economic, and environmental impact
Customer: Higher product/service quality at a lower cost
Employee: Contribution, dignity, and quality of life

Currently safety is not directly addressed in either of the ISO standards. However, although ISO 14000 does not specifically refer to occupational health and safety, ISO 9000's preamble suggests that safety could be a consideration in the establishment of the QMS's. In the United Kingdom, Ireland, Australia, and New Zealand, the health and safety regulatory authorities are establishing health and safety management systems built on the ISO management system's standards. In addition there is a draft Occupational Health and Safety management system standard 18001.

Because ISO 9000 and ISO 14000 draw from the same quality principals taught by Deming and Juran, it is relatively easy to integrate them. These systems make it much easier for multi-site and, more important, multi-national companies to consistently execute effective EH&S and QMS's. As you can see in Table 2 the organization and emphasis on the various management systems is different, but it fundamentally strives for process control and conformance by using the same management tools.

Today many companies around the world have implemented ISO 9001–14001 management systems for quality and environmental management. If you read these standards or any number of technical references describing them you will quickly see that they have their roots buried deeply in the quality movement started by Deming and Juran. ISO aims to standardize and integrate disparate regulatory and best practice systems and to reduce duplication of effort.

Health and safety is only one of the many responsibilities of managers striving to compete in their marketplace. Therefore, we must do everything possible to leverage efficiency and, more important, prevent losses. QEH&S management systems must ensure current performance and continual improvement.

The International Labor Organization estimates that there are now 37,000 multi-national corporations and 150,000 facilities with more than 70 million

TABLE 2. ISO Registration Trends

ISO Registration
ISO 14000 Registration
• Japan >500 sites
• Korea >65 sites
• United States 90 sites, in 32 states
ISO 9000 Registration
• 1993 ~ 20,000 sites
• 1995 ~ 70,000 sites
• 1996 ~ 170,000 sites

employees, 22 million outside their parent company's country. Thus, the need for standardized and IMS's is confirmed. Companies cannot afford the redundancy and its associated costs caused by multiple and fragmented systems around the globe. In addition, most corporate council will advise their boards that inconsistent implementation of policy will create potential liability for them in the event of a serious accident. In Table 3, integrated process management goals give managers a simple overview of the goals, processes, and tools that will be needed or evaluated when implementing any management system, but should be of particular interest when integrating a QEH&S Management System.

It can be reasonably expected that more companies will integrate their QEH&S programs as they become more familiar with their ISO management systems and see the opportunity to more efficiently include work place safety as an integral element of their system. Internationally, regulators and corporations are regarding the American (U.S. OSHA) prescriptive approach to standard enforcement, although necessary in some cases, as not as effective as performance improvement based on a management system that allows managers to better understand and control process risks. A recommendation that I will make repeatedly in this text is that your management team should integrate QEH&S management systems.

TABLE 3. Integrated Process Management Goals — Tools

Harmonized System Goals

Do it right the first time and avoid rework, waste, or accidents
Establish standards and defined work procedures
Identify and document control of hazards
Measure both direct and indirect cost of process nonconformance
ID critical tasks and internal and external customer needs, including their definition of quality
Understand process risk
Facilitate process improvement teams involving all levels of operations
Supervisor — management job observations and coaching
Nonconformance multiple root cause analysis, and sharing of lessons learned

Key Considerations or Elements of Process Control	*Essential Process Management Tools*
Personnel selection, training, and supervision	Inventories of critical tasks, processes, equipment, and tools
Engineering	Risk assessments
Materials and procurement	Process observation/analysis, check sheets
Maintenance (facility, equipment, and tools)	Audits, surveys and inspections (process monitoring and evaluation)
Procedures/standards	Accident and process nonconformance investigation
Management and supervision	Corrective/preventive action plans, including communications and aggressive follow-up
Communications	Accountability: budget and charge back to the work center for process failure, rework, environmental liability, and accidents (cost of nonconformance, both direct and indirect costs)

To do otherwise would mean a significant duplication in resources and at best a weakly coordinated process improvement effort. I remind clients to look at this situation from the perspective of their internal customer, their supervisors, and their work center employees. In many organizations the supervisor is held accountable for budget and schedule goals with weak or non-existent focus on EH&S and a moderate focus on quality. One of the primary difficulties created by the fragmentation of these process management systems is competition for scarce resources and management or involvement in the systems. Process control fragmentation ultimately creates three different teams (safety, quality, environmental) evaluating critical work processes. Each of these separate process control teams requires the supervisors' and work centers' time commitments for process improvement and support of systems for QEH&S risk reduction. This requirement results in poor utilization and ineffective or duplicated efforts of facility resources. Many times this inefficiency frustrates supervisors who feel the pressure of budget and schedule demands, and it results in passive support or in some cases malicious compliance for processes they don't believe directly support this focus.

Most organizations need to address these questions as part of their quality or EH&S planning efforts. What do the supervisor and manager need? How can we make EH&S a seamless component of process management and quality control? What can employees do to reduce rework, waste, duplication of material or package handling, package damage or loss, and injuries? What tools are needed? What are the life cycle risks associated with a product, including the cost of post use disposal?

Far too often management does not understand who its internal customers are. Identifying your external customer as the person or company purchasing your products and services is easy. Oftentimes managers do not as easily recognize internal customers. I believe that this lack of recognition is largely due to internal competition and a tendency to not step out of self-imposed definitions of what your job is versus trying to do what is needed. The first line in any position description should require employees to responsibly participate as team members performing whatever task their work center or the facility management needs to service customers.

This problem is also exacerbated by labor organizations' trade-protection activities and not allowing members to cooperatively support each other across trade lines. Too often this type of self-imposed inefficiency has lead to the closing of facilities and in some cases entire companies. Protectionist attitudes are generally rooted in fear or a desire for power and at best satisfy individuals short-term personal goals, but in the process these attitudes sacrifice the good of the members and the company. I have used the model in Figure 3 for years to help managers and supervisors understand process management and the critical link between the quality of process design and process input with the quality of the process outcome. As the old adage goes, "Garbage in equals garbage out!" The quality of process inputs also goes to my argument for integrated risk or process management. Without it, companies will not only experience additional product nonconformance or rework but, as can be seen in the model below, they will also experience additional injuries and environmental liability. If all of the process inputs are not appropriate or are incorrect in any way, the risk of process nonconformance is increased according to the level of faults or risks allowed in the process-input design. Therefore, any reasonably designed QMS must consider the quality of process inputs and how to standardize the process and the measurement of system controls.

Figure 3 illustrates the importance of ensuring that the primary process inputs are correct. If they are sub-standard, then sub-standard work behaviors, rework, loss of product, injuries, or environmental liabilities result. These nonconformances will manifest themselves often as minor incidents or nonconformances. In some cases these less serious incidents will occur hundreds of times before a serious process failure results in employee injury or death, loss of customers, or significant environmental damage.

Process Management

- **Performance Standard:** Schedule, budget, customer service, or product requirements.

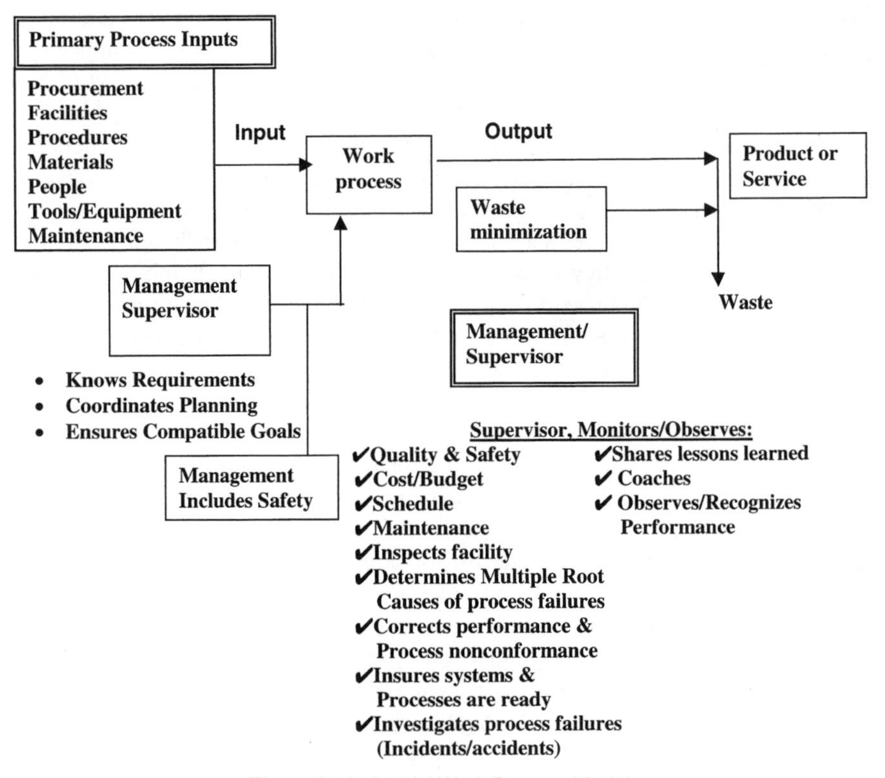

Figure 3 *Industrial Work Process Model.*

- **Support for the Process:** Process specifications; Trained/qualified personnel and management/supervisors; Tools/equipment, including maintenance and software.
- **Prevention:** Clear communications and job requirements are needed as well as well-designed processes, policies, and systems.

Companies cannot risk the adverse public relations and corresponding loss in market share by sacrificing the safety and environmental health of their community and employees. This belief is reflected very strongly in the chemical manufacturer's responsible care (CMRC) program. However, quality management or EH&S at any cost will not be tolerated. Smart and practical solutions are expected. Companies want and need to move from a compliance-regulatory approach to process sustainability and systems management that cost-effectively improves the process.

Multi-national corporations are demanding consistent global facility management, and at the same time they expect their organization to improve its competitive position in the marketplace. Thoughtfully designed systems or process management, including the integration of QEH&S, can help companies realize this goal.

It can be reasonably expected that more companies will integrate their QEH&S programs as they become more familiar with their ISO management systems and see the opportunity to more efficiently include workplace safety as an integral element of the system. The big three automobile manufacturers' QS 9000 QMS based on ISO 9000 illustrates how seriously some large companies are taking ISO management systems. This standard is required for all vendors providing parts and services to these manufactures. In addition Ford and GM both announced in December 1999 that vendors will be required to comply with ISO 14001. As mentioned earlier there is also the draft 18001 standard for OH&S.

The EPA recently announced it would promote environmental management systems, including ISO 14001. The new proposed OSHA standard also has very similar systems goals or management elements; these processes include:

- A goal of improving work processes, doing it right the first time, and eliminating rework, waste, down time, and other impediments to productivity — continuous improvement.
- Measuring (bench marking) the frequency and severity costs of nonconformance as compared with other industries or facilities, performance standards, total costs, and trends by work center, against measurements of expected performance.
- Establishing teams to evaluate and design process improvements or to perform process risk assessments (quality action or process enhancement teams/safety committees).
- Identifying critical tasks and evaluating the risk associated with the process.
- Evaluating the opportunity for process improvement or re-design. Tools such as job task analysis, performance observations measurements, systems safety reviews, or, if data are available, loss source analysis.

- Clearly identifying internal and external customers and understanding their respective needs.
- Setting standards and defined procedures for work performance.
- Analyzing process failure (multiple root cause analysis).
- Measuring baseline cost data for both direct and indirect costs of process failures, accidents/incidents (all are process nonconformances).
- Implementing effective corrective/preventive action plans with active follow-up.
- Providing supervisor coaching, correcting substandard performance, and commending desired performance (reinforcement) (Note: This approach requires a commitment to supervisors' and management training or development).
- Requiring that internal and external customers provide input in defining quality and performance measurements.

Process Improvement/Risk Assessment. To ensure quality and EH&S, the process must be defined to identify improvement opportunities and risks or potential deviations that could threaten conformance and safety. Employee involvement in the process review is critical. Employees oftentimes are the true experts on the process, and they know the process risks or deviations that compromise quality and EH&S. Employee involvement can include participation in quality circles, Process Enhancement Teams (PET), safety committees, facilitated risk assessments, and performance monitoring and job observations. Multiple root causes of process nonconformance must be identified if process risks are going to be eliminated. Injuries and rework causation are oftentimes caused by the same multiple root causes as quality or environmental process failure. You don't know what you don't know, and an excellent way to bring clarity to the process is to utilize your most valuable resource's creative energy!

Managers must believe that most employees want to do the job right but are oftentimes confronted with system or process problems owned by management. We accomplish process improvement by evaluating systems/processes with tools such as task analysis, process flow diagrams, systems safety engineering, check sheets, critiques, process design, and training. Work center involvement and group participation, including employee participation in quality action teams or safety committees, are as important as senior management involvement. In many cases, it can be argued that these factors are more critical, especially when evaluating the risk associated with high-hazard operations. Employee and management involvement—and more important, commitment that includes allocating resources—are absolutely necessary. You cannot leverage to improve process design at the expense of employees or supervisors, making them wrong for system or process nonconformance. Although an employee may have done something to contribute to the process failure, what was it about the process that allowed or caused this failure to happen? What were the substandard conditions or common causes contributing to 85% of the nonconformance issues?

Too often incident or process nonconformance investigation reports find that the cause of the system failure was employee carelessness. Many times these investigations recommend corrective actions that include admonishment of the employee and additional training, failing to recognize the multiple root causes of the incident, or process nonconformance. Therefore the substandard primary process inputs (see Figure 1) are not corrected, leading to additional process nonconformance and costly incidents. As a manager, ask yourself how many employees would come to work with a plan to injure or kill themselves? Therefore, what were the multiple root causes of the nonconformance or accident? What changed and what is in the management system or work environment that causes substandard performance or other process deviations and nonconformances? How can we improve the production or operations process? Employees and managers must have training that includes process design, procedures, and sharing lessons learned from nonconformance evaluations, incidents, or risk assessments. Unless a systems approach to process management is adopted, the multiple root causes of process failure that exist in the management system and facility work environment will not be recognized, which will lead to more serious incidents or process failures in the future. Many times employees and supervisors introduce changes, small process deviations to expedite the process or overcome an existing obstacle or process problem (work arounds), without evaluating the risk associated with the change. Over time these deviations and others like them result in significant process failures and incidents.

Unless there is fraud or sabotage, most employees want to do the job right. However, management may have failed to give employees the tools or resources to meet their expectations, failed to appropriately communicate needs, and created an environment with conflicting requirements, communications, or expectations. Because supervisors tend not to evaluate or observe jobs frequently enough or to really think about the activities they see (e.g., questioning the activities), they allow small changes to gradually replace Standard Operating Procedures (SOP)s. Note the difference between observing and thinking about a job and simply seeing it. These small well-intended changes compromise the process and ultimately result in a process failure, including serious injuries, quality and environmental issues. The bottom line is that management owns the process and must therefore be accountable for its outcomes (good or bad). Management must take time to observe jobs and to evaluate the job or execution of the task to ensure that it is being done in accordance with requirements. Managers must also take time during these job observations to ask the critical "What if" questions to better understand process risk and the adequacy of or need for additional controls. Performance or process behavioral observations are a critical aspect of performance measurement and reinforcement. All of the management systems require managers to define, understand, and continuously improve process management, including EH&S. A process can be analyzed only so many ways, and it is infinitely more effective to integrate process management and to focus on continuous improvement without treating EH&S as a separate subject. Besides, how can you have quality when you are injuring people or causing the environmental problems?

The emphasis of ISO harmonized process management systems is an attempt to develop controls that do not rely excessively on the behavior of people. More emphasis is placed on process management or systems safety, ensuring that the work environment and all of its critical elements outlined in Figure 3 are designed and executed in a way that ensures process conformance. In addition, an underlying expectation is that in the event of nonconformance or process failure, managers will identify the multiple root causes of the failure or accident before developing corrective and preventive actions.

Process Control Facts (Ref. 22)

Definition: A logical decision can be made only if the real problem is first defined.

Multiple Causes: Problems are seldom if ever the result of a single cause.

Critical Few: In any given group of occurrences, a small number of causes will tend to give rise to the largest proportion of results (process success or process failure), "Parieto Effect".

Point of Control: Process control potential lies with the first-line supervisor, however, senior management must support the supervisor with good work procedures, tools, materials, etc.

The first step in integrating management systems is to complete an evaluation of basic process risk factors and an independent audit (gap analysis), and possibly a corporate-climate culture review. This step is necessary for management to understand clearly and to plan for system implementation. Management must also understand that the audit and management system implementation requires personal commitment, including resources (financial, material, management, and staff resources). ISO harmonized management systems are applicable to any organization that wants to improve QEH&S process performance, regardless of size, type, or level of maturity.

The initial management systems audit compares the current situation within the company with each element of the soon-to-be-implemented or recommended QEH&S management system. Audits recognize significant failures or process nonconformances and will make corresponding system recommendations to deal with a system of non- or partial conformance.

Initial system and periodic audits by independent third parties or cross facility audits must be performed so facility managers can stay on top of their performance. These audits must also include a follow-up or status review of action items previously agreed to as part of the IMS implementation or continuous improvement process. Management policy regarding EH&S must be consistent with process control/quality requirements. Project, facility, and product design, or production plans must include systems safety integrated with the quality management design or planning function, including product life cycle risk analysis and mitigation.

Programs must begin with an evaluation of total risk. Managers should perform risk assessments and map risks to prioritize quality and EH&S activities.

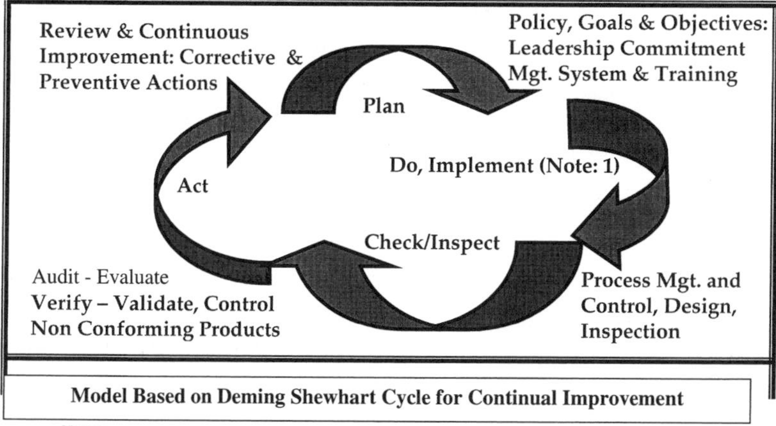

Note 1:

- Structure, Responsibility
- Training, Awareness Competence
- Communications
- Documentation & Document Control
- Design Control
- Operational Control
- Purchasing Control, including management of contractors
- Emergency Response & Contingency Planning
- Mgt. Review/Evaluation

Figure 4 *Continuous Improvement.*

Conducting climate culture reviews can also be very useful. The climate culture review enables management to validate the effectiveness of system implementation as seen through the eyes of the employees and the various management levels. It injects a strong dose of reality into the management system implementation effort. Figure 4 is a simple expansion of the Deming–Shewart continuous improvement model for quality. One of the important illustrations provided by the model is that it recognizes that time and commitment are required to evaluate and identify process risk and to take appropriate risk-ranked corrective actions. Management must then reassess the impacts of the changes before planning the next phase of process improvement. This continuous improvement process (plan-do-check-act) repeated over time creates a never-ending journey to zero defects.

Table 4 (below) further illustrates the Deming–Shewart model design in the chemical management or the environmental life cycle analysis of production processes.

Integrated Process Quality and EH&S (QEH&S) Program Benefits

- Reduced costs increase profit margins
- Increased competitiveness
- Facilitates injured employees to return to work
- Reduces incident frequency and severity rates or lost time

TABLE 4. Life Cycle Continuous Improvement Considerations

Plan	*Do*
Each product life cycle and related production, distribution, use, and disposal process, including policies and impacts to stakeholders, is assessed by QEH&S, product design, and work center experts (systems safety–life cycle risk assessment)	Waste, rework, environmental releases, injuries, and customer complaints are evaluated to determine the multiple root causes of the process nonconformance
Act	*Check*
• Establish policies, procedures, and standards of practice • Train responsible managers, supervisors, and employees on the process execution goals and documentation procedures • Share experience (lessons learned) with other work centers and facilities • Follow-up on audit/process improvement recommendations to ensure effective implementation	• Measurements and observations are evaluated by using statistical tools • Every process nonconformance is reported and evaluated • QEH&S audits are performed to assess the effectiveness of the management system (annual internal audits and two-to-three-year external or third-party audits)

- Reduces damage to equipment, inventory or product loss, and generation of hazardous waste
- Increases companies' regulatory compliance
- Integrates process quality and safety
- Improves employee and public relations

Milliken Corporation, Dupont, Alcoa, and Deere & Co. are all good examples of how focused leadership can successfully reduce losses in EH&S, while enhancing quality and shareholder return. Milliken is the largest privately held textile company in the world, a Malcolm Baldrige Award winner and a company with a very successful QEH&S record, starting with strong personal commitment from Milliken himself.

In DuPont's 1997 Safety, Health and the Environment Progress Report, John Krol, DuPont's CEO, stated, "Employee injuries and illnesses are down 27% globally since 1996, and 60% since the early 1990s. DuPont's performance continues to be five times safer than the chemical and petroleum industry average, and 20 times better than the average for all industry. Also, our contract partners improved their safety performance by 45% in the last three years. Total release to air, land, water, and underground injection wells in 1996 were 20% lower than in 1995 and almost 80% lower than in 1987. By the year 2000, we expect TRI releases to be down another 50%, or more than 90% from the 1987 base year."

During this same period DuPont's total shareholder return increased 300%! In the 1992 annual report, Deere and Co. reported that "Deere continues to be a leader among U.S. industries in occupational safety. Fifteen John Deere factory, marketing, and administrative facilities won National Safety Council performance awards... Employee injury and illness rates have been reduced by 93% since 1975, when the company's occupational safety program was put in place. This represents an estimated $263 million in workers compensation cost avoidance during this 17-year period—$35.8 million in 1992 alone."

As discussed earlier, supervisors tend to be more concerned about their production schedules and budget. They oftentimes fail to recognize the opportunity to enhance both of these goals by having an effective process or systems safety program that integrates their quality and environmental management programs. Throughout this text we will refer to systems safety. However, you must understand that the various assessment and control tools used with the systems safety are quite broad and are not just for personal safety. Systems safety is an organized way to evaluate systems risk and process readiness. It therefore has value in both quality and environmental process evaluations and development of process controls. Program managers of various processes control systems, such as safety, quality, and environmental, too often confront supervisors from their specialized process perspective or niche only. Managers analyze and evaluate the supervisor's process from their biased expertise. The ultimate result is that the supervisor becomes frustrated and returns to focusing on more immediate and certain pressures such as budget and schedule. Process control fragmentation caused by disjointed QEH&S program implementation will tend to cause supervisor overload and weak commitment to QEH&S. This tendency is especially true when the supervisors have not received adequate training. Training of all management and employee levels must be supported and should include technical, system, and managerial training. The supervisor is a very important element of process control and is in effect the gatekeeper for process control management. The supervisor is both the internal and external provider of customer service that includes employees.

The supervisor has a critical role that should receive much more attention and management support than it currently receives in many companies. Inherent in this comment is a belief that most companies need to do much more to train and prepare supervisors for their positions and to develop individuals to be the future leaders for the company.

Now that I have emphasized my biases for supervisors, please don't think that this means that senior management is not equally important. Senior managers are obviously critical, but in a different way. They should provide leadership and develop supervisors as their customers to execute the corporate mission and policies on quality and EH&S. They must also ensure that the primary process inputs provided to employees are not substandard if they expect to avoid incidents. If senior management is not committed and involved, including ensuring that the supervisor has needed resources, the process will quickly fail. In addition, performance measurements will not be appropriately focused and can cause inconsistent or conflicting messages to be delivered to the people on the production floor. This scenario will soon cause the focus to shift back to that which management is serious about measuring (oftentimes budget and schedule). Also, it will contribute to undesired behaviors associated with QEH&S. This cycle causes injuries or quality process incidents and ultimately impacts the schedule and budget that triggered the change in process performance. A vicious-reactive cycle at the opposite end of the Deming–Shewart process improvement model.

Corporate Quality and Sustainability Challenge

- Who will lead?
- What is the inspiring integrated vision?
- Where will the sense of urgency come from?

If a process is failing as reflected in rework, damage to equipment, products, or injuries, it needs to be evaluated and re-defined to eliminate or reduce the potential for future process failures. The multiple root causes of the nonconformance must be understood before corrective and preventive actions can be effectively implemented. Effective and thorough process evaluations or risk assessments can only be accomplished through joint employee and supervisor cooperation, including participation in critical job performance observations and risk analysis.

Proposals should as a minimum include ROI estimates, discounted cash flow estimates, and/or use simple risk management tools such as Microsoft's Risk Map Tool. We must communicate to management in terms that they understand and can integrate into their business decision process. Too often, EH&S managers are upset because their recommendations are not acted on. However, too often they present their proposals supported by subjective data and very little credible financial information.

Tools such as Risk Map and the Liberty Mutual Savings Equation allow you to map the severity and frequency of process or facility risk. To use these tools you must define your project costs, anticipated loss reduction, internal rate of return, depreciation rate, and inflation rates. The resulting reports are short easy-to-understand financial arguments for control options.

Without learning how to take advantage of these types of tools the process control or QEH&S manager will find it increasingly difficult to compete with other capital or management system requests for financial support or corporate resources. However, key to successful implementation of this type of analysis is well-defined assumptions or data regarding the costs of nonconformances or accidents, and estimates regarding the effectiveness of loss reduction expected over time for proposed controls.

Management review and continuous improvement of all aspects of the IMS will require a commitment of resources to ensure that they have the information necessary for process improvement. In the end, we must all be willing to back away from our professional "rice bowls" and recognize that a process is a process is a process, and only integrated QEH&S management will ensure sustained corporate success.

Integrated QEH&S Risk Management Model Objectives. The basic objective of an integrated system is the control of accidental losses and costly process nonconformance by taking the following steps.

- Identifying critical tasks and evaluating and ranking their risk for possible process improvement or redesign.
- Clearly identifying internal and external customers and understanding the clients' needs.
- Establishing teams to evaluate and design process improvements (PET).
- Establishing SOPs for work performance.
- Analyzing process failure (task analysis, root cause, and multiple cause analysis).
- Measuring performance and evaluating performance outcomes, including the cost of nonconformance.
- Developing baseline cost data for direct and indirect incident/accident losses (process failures).
- Developing and executing effective corrective action plans.
- Conducting training, including sharing lessons learned and process change briefings.
- Correcting sub-standard performance and commending desired performance (behavioral-based performance management recognition).

In Singhal, 1999, Dr. Singhal writes, "I have conducted an extensive study on the financial performance of companies practicing TQM. Our key finding is that although TQM is not a quick fix, it does pay off significantly. For example, the stock price of companies with effective TQM implementation out performed S&P 500 Index by about 34% over a five-year period."

- In GE's 1997 annual report, they stated that Six Sigma delivered more than $300 million to its operating income and in 1998 they expected to double that. Other savings discussed in Reference 14 include these examples.
- GE's Medical Systems Group described a 10-fold increase in CT Scanner Life, increasing "up time"/machine, profitability, and improved patient care.
- Superabrasives described how Six Sigma quadrupled ROI by improving yields.
- Railcar businesses had a 62% reduction in turnaround time at its repair shop, two to three times faster than its nearest rival.
- The plastics business increased capacity by 300 million pounds — equivalent to a new plant or $400 million in savings.
- Average Black Belts save the company $1 million a year.

- GE anticipates $10–15 billion in increases in annual revenues and cost savings by 2000.
- Raytheon estimates that 25% of each sales dollar goes towards fixing problems and that Six Sigma will reduce this figure to 1%.
- Allied Signal increased capacity for Caperolactum (used to make nylon) at savings of $30–50 million/year, avoiding the need to construct an $85 million plant.

SUMMARY

All of the efforts discussed in this chapter require management to define, understand, and continuously improve process management, including EH&S! Practical solutions are needed that merge quality and EH&S. Programs must look at total risk and be integrated with facility management systems. Too often we tend to focus on meaningless documentation exercises or on regulatory compliance versus process performance, which will make these effort less difficult or eliminate them, especially the regulatory impact.

"A recent study concluded that 88% of the money paid out between 1986–1989 by insurers for Superfund claims went to pay for legal and administrative costs, while only 12% was used for actual site cleanup" (Acton & Dixon, 1992). It generally has been agreed that too much of the money paid for environmental costs in our society by business and government is spent on regulation, enforcement, and various other legal and transaction costs and not enough on actual cleanup. Recent cleanup projections have shown it can be far more expensive to clean up a polluted body of water or improve an old, poorly conceived landfill than to apply precautionary measures to prevent problems in advance. Active and competent environmental protection is just as important to Ciba's (Ciba-Geigy) survival as the conventional pillars of research, production, and marketing." (Epstein 1996).

Quality and good process management depend on people who respect each other and have pride in their company. The company's management must have integrity, and it should understand the cost of nonconformance and how to satisfy its customers' (internal and external) needs. Customers will always view value as a service or product that is as good or better than what they expect from your competition and is provided at a competitive price that is not necessarily the lowest. Crosby defines the price of quality as the costs of doing things wrong and the price of ensuring things are done right the first time. Crosby's quality absolutes are

- **Definition:** Conformance to requirements
- **System:** Prevention
- **Performance Standard:** Zero defects
- **Measurement:** Price of non-conformance

Integrating environmental and safety management systems into an overall process management or QMS makes good business sense. You cannot "shoot" for zero defects as the companies implementing Six Sigma are without designing out EH&S risks and liabilities. Peter Drucker (1998) states, "Economists have been wont to consider pollution and environmental damages as externalities." The costs are borne by the entire community rather than by the activity itself. But this situation will no longer do for environmental damage. There is no incentive not to pollute. On the contrary, to pollute without paying for it confers a distinct competitive advantage on those who pollute the worst. To treat environmental impacts as "externalities" can also no longer be justified theoretically. During the past century every developed country has converted industrial accidents from an externality into a direct cost of doing business. Every developed country has adopted workmen's compensation under which the employer pays an insurance (risk transfer) based on its own accident experience, which makes the damage done by unsafe operations a direct cost of doing business. Workmen's compensation assumes that industrial activity is inherently dangerous; accidents are thus bound to happen. This assumption, regarded by some as a "license to kill," was bitterly fought by reformers dedicated to making the workplace safe. Workmen's compensation has done more, however, to reduce the incidence of industrial accidents than safety regulations or factory inspection."

IMS Pros and Cons. One of the key considerations is that the goals of an OH&S, EH&S, or QMS are all the same. All three are trying to establish a system of performance standards, and first time/every time conformance to those standards, thus ensuring customer and stakeholder satisfaction, maximization of profits, and minimal process risk, or less. Organizations with effective IMS should be better able to control process risk and ensure customer satisfaction. Decision-making will be improved, and the level of management participation in identifying and controlling process and production risks significantly enhanced.

An IMS offers organizations the opportunity to improve among other thing business effectiveness, as well as QEH&S performance. The process of implementing a system, however, will require significant management commitment and organizational effort. Table 5 summarizes some of the more common pros and cons for integrating management systems.

ACTION PLAN CONSIDERATIONS

At the end of each chapter I list things for you to consider or do when thinking about implementing your management enhancement system.

- Do you have a QMS? If so have you considered integrating it with EH&S?
- Does operation's management meet with quality, risk, and EH&S management to evaluate process risk and jointly plan process improvement?

TABLE 5. Pros and Cons for IMS

PROs	*CONs*
Ultimately more cost-effective	IMS if not managed properly could become over-centralized and complex
Improved product and production life cycle plan, decision making, and risk management	There will be some organizational vulnerability and risk during the organizational change process, for example, systems not fully in place, and reviews or process updates may slip
Objectives and process management systems are basically the same	Safety and environment management systems, unlike quality, can be heavily influenced by regulations
Avoids duplication of resources and unintended process control designs that may be in conflict, when designed in the specialty vacuum of quality, environmental, or safety	Regulators and third-party auditors may initially find it more difficult to evaluate the IMS when it is quite properly integrated into the overall business system
Integration should prevent possibility of resolving problems in one process risk area while adding to process risk in an other area	
Quality procedural control instructions may be developed more easily, including environmental and safety risk controls	
Improve management review and organizational risk management process	
Improve communications and sharing of process control and risk assessment techniques between quality, operations/production, environmental, and safety disciplines	
IMS should result in better risk ranking and commitment of organizational resources	

- How are the total costs of process nonconformance and incidents identified?
- Does your company consider life cycle risk in its product design and product planning effort?
- How are process nonconformances evaluated?
- Do employees and supervisors participate in risk assessments or process improvement efforts?
- Does the company or facility have a prioritized list of process/facility risks and associated action plans to eliminate or mitigate risk?
- How does the company benchmark QEH&S?
- Have you implemented consistent EH&S policy and procedures (best practices) at all of your facilities, including global locations?
- When presenting management with recommendations for QEH&S process improvement are all costs (direct and indirect) estimated with clearly defined assumptions, and are the expected benefits of various control alternatives also presented, again, with well-defined assumptions supporting the benefit statement?

- How do you plan to work with other managers, including operations and finance, to ensure that process improvement practice is consistently applied in your organization?

The process flow diagram below is an overview of what the organization must consider when implementing an integrated management system. It is really the road map for evaluating and considering needed organizational change that will be used by the executive management QEH&S Steering Committee. No single

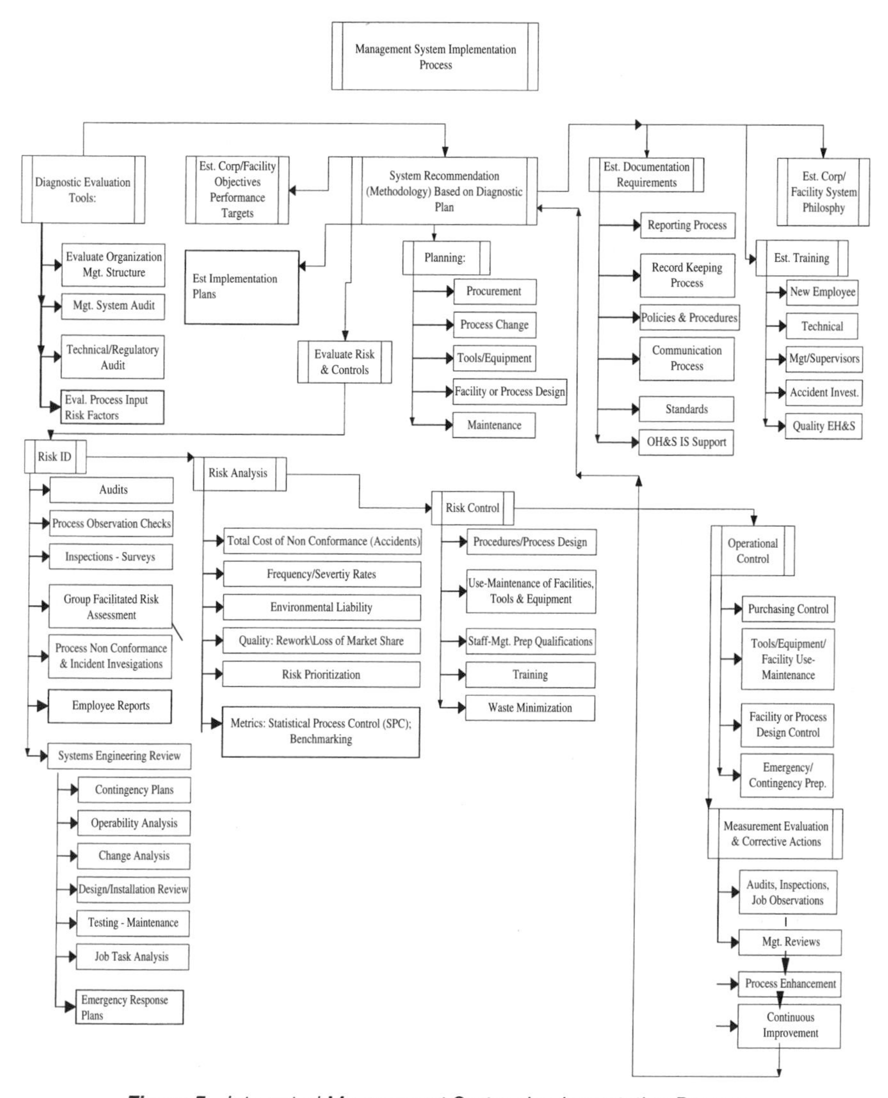

Figure 5 *Integrated Management System Implementation Process.*

department, risk assessment team, or PET will be responsible for everything identified in this process.

I see implementation being executed by taking advantage of the exchange of skills and experience enjoyed in matrix-type organizations.

2

Building Safer More Productive Jobs

The cost of accidents in 1992 equaled 31% of all pre-tax operating income and 78% of shareholder dividends.

Most employers have become increasingly concerned with the costs of workers' compensation and health benefits. The direct losses due to injuries, illnesses and process failures are often just a fraction of the actual losses/experienced by a company. Studies of accidents and process failures also document that there are oftentimes many warnings in the form of process rework, minor injuries or near misses before there is a serious process failure or fatality. This chapter focuses on safety as a key element of process improvement and quality enhancement. To me, it is almost unimaginable for companies to implement quality management systems (QMS's) and not include safety in the process improvement effort.

Figure 6 illustrates the ratio of minor incidents to serious accidents and nonconformances. These early warning signs of substandard processes are too often ignored because management has not thought about how the outcome of the minor incident could have been much more serious, given slightly different circumstances.

Ten years ago California implemented legislation that attempted to require some elements of a process management system for safety. California's Occupational Safety and Health Administration (OSHA) adopted legislation (Senator Bill Greene's Accident Injury Program, Legislation SB-198) on October 2, 1989, requiring management to evaluate processes, perform risk assessments, and develop management plans to mitigate the identified risks. OSHA's proposed safety management standard and VPP programs reflect a growing understanding that QMS's approaches to safety is much more effective than prescriptive-type

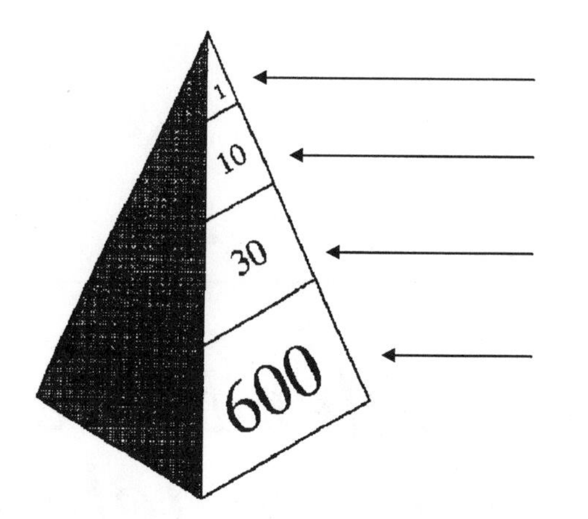

Figure 6 *Accident Ratios.*

regulations. In addition, countries around the world such as Australia and New Zealand have adopted an occupational health and safety (OH&S) management systems based on the United Kingdom's BS 8800 standard. We also have a draft occupational health and safety (OH&S) management system standard that was developed with input from several counties and organizations using the BS 8800 standard as a model. This draft could ultimately become the model for, and the ISO standard or elements of it could be incorporated into an IMS Standard. (OSHAS 18001)

California's Injury-Illness Prevention Program (SB-198) basically requires the following specifications.

- Written progress reports
- Designation of a responsible manager, preferably the senior executive of the facility
- Performance of job-taskjob-task analysis and/or systems safety reviews
- Development of job/process procedural controls and a written program
- Development a system and executing a process for periodic in-house inspections (including check lists)
- Development of systems for accident/incident investigation (provide procedures, including forms and training of managers on the system)
- Establishing employee communications systems, for example, tailgate meeting, committees
- Record-keeping, documentation of all of the above

Unfortunately this well-intended law consisted of a number of "fill-in-the-blank" paper programs for complying with this OSHA requirement, which fail to seriously evaluate and control process hazards.

European Union requirements mandate management audits or risk assessments of facility programs, performance of risk assessments, and ultimately the design of programs to reduce or eliminate risk. In Europe the regulators expect companies to evaluate their facility and process design and to control their processes through well-thought-out risk assessments and mitigation plans that can form the foundation of an OH&S management system. In Australia, some states require self-insured companies to implement a process called safety map, which is harmonized with ISO 9000.

OSHA is promulgating an OH&S programs standard 29 CFR 1910.700 with provisions similar to OSHA's voluntary guidelines. The standard requires employers to identify and control work-site hazards and to involve employees in all phases of the program. Marthe Kent, Director of OSHA's Office of Regulatory Analysis, stated that "If OSHA had it to do over again, this would probably be the first standard we would promulgate because it truly provides employers and employees with the foundation for workplace safety and health...our evidence suggests that companies that implement effective safety and health programs can expect reductions of 20% or greater in their injury and illness rates and a return of \$4–\$6 for every \$1 invested."

Companies with ISO 9000 or ISO 14000 programs are starting to evaluate harmonizing worker safety and health into these programs and, to the extent possible, integrating the management aspects of these quality and environmental management programs.

Listed below is a table outlining the common elements of most QMS's, including those of Deming, Juran, Crosby, and—more recently—Total Quality Management (TQM). Table 6 also compares some of the major elements of most QMS's, with similar requirements in any well-developed OH&S Management System.

In the United Kingdom, the chemical and rail transport industries are required to develop a management system referred to as a safety case, a detailed cradle-to-grave management plan for handling hazardous materials, or executing high-hazard operations. The safety case requires that the facility perform risk assessments of all hazardous operations and develop plans to mitigate the defined risk. Plans must also include public communication and emergency or contingency plans. The safety case requires:

- Risk assessments
- Management system policy with clear objectives
- Auditing, monitoring (could include third-party inspections and planned job observations to measure compliance with critical task requirements)
- Accountability and follow-up
- Non-conformance accident investigation process
- Controls: operational and maintenance
- Contingency-emergency response plans and drills
- Implementation plan, including followup and evaluation

TABLE 6. Quality Management TQM and OH&S Process Management

Quality	OH&S and Environmental
Management commitment and organizational constancy of purpose	Same
Adopt new management philosophy, including Crosby's Four Absolutes: • Quality is conformance to requirements (defined by the customer, internal and external, and defined in process requirements, training, etc.) • The system of quality is prevention • The performance standard is zero defects, 100% conformance to defined standard or process requirements • The measurement of quality is the price of nonconformance (direct and indirect costs) and frequency rates, the ratio of conformance to critical job/task procedures	Same
Must have a program of continuous improvement	Same
Institute on-the-job training (OJT) and rigorous program of education and self-improvement for process improvement, which includes quality, safety, and environmental requirements	Same
Leadership: Create open, trusting environment	Same
Drive out fear	Same
Break down barriers between departments and internal customers	Same
Understand and implement process for error-cause removal, identify the multiple root causes of process failure and accidents	Same
Employee involvement: Quality action or improvement teams, critical job observations	Safety committee, facilitated group risk assessment, performance job observations, or quality improvement team
Support results through recognition, positive-immediate feedback, and coaching, and correct undesirable behaviors to ensure process standards or requirements are executed as intended	Same

- Training and communication
- Management evaluation

As you can see the requirements outlined here are very similar to what is required in ISO management systems and TQM. The lesson to be learned is that by integrating their quality and environmental health and safety (EH&S) systems, management will significantly leverage the effectiveness of its efforts and resource utilization.

Understanding "process management" and "systems safety" are critical to identifying process or business risk. I will once again emphasize that systems safety is a process for understanding work processes and that it establishes a system for identifying the risks associated with that process. Please note that this emphasis does not mean that systems safety is strictly a safety tool, it is a management tool, and does not and should not be considered a tool only for the corporate safety engineer or safety consultant. Operations managers must manage the process and rely on the EH&S managers or technical consultants; systems safety can be an invaluable tool for helping them understand their processes or system risks. Experience performing only OSHA-type regulatory audits will not be enough to qualify an auditor to perform competent management system audits. In addition, the use of audits is critical to understanding the level of systems, process, or program conformance. To be successful, however, an auditor must understand process management, how to ask the right questions, and how or what to appropriately review as documentation. Your best auditors will have good technical and managerial skills.

Companies like DuPont, which has made safety part of its corporate culture since they first began making gunpowder, clearly demonstrate the benefits of managing quality and EH&S. DuPont's continuing efforts to reduce its already remarkably low injury illness rate is also matched by increases in productivity and the shareholders' stock value. Clearly, this documentation should make other companies question the argument that EH&S is too costly and therefore cannot be afforded. Good EH&S is good business! Companies like DuPont see no value in fragmenting work processes. These companies regard process changes that improve quality as an opportunity to improve safety and environmental performance. Over the past 10 years they have increased shareholder equity by 300%, while reducing environmental liability and injury illness rates by 60%. The savings generated by improving process management freed revenues that clearly went straight to the bottom line.

From a more simplistic perspective, think of the efficiency and increased cooperation that can be realized among supervisors by discontinuing the fragmentation of process management into quality, environmental, and safety divisions. After 25 years I have never investigated or reviewed an investigation of a serious process failure, injury, or fatality that did not involve multiple root causes in the management system. The investigations also demonstrated such an incident result, and at times generated environmental liability. The direct and indirect costs of these related impacts are many times not well-documented and therefore do not attract the management attention they deserve.

I hope this book establishes better understanding of these issues and the need to integrate the various management systems bombarding the workplace. Because ISO 9000 seems to be gaining in global recognition, we will also focus on it and 14000 as the foundation for beginning the process of developing a sustained and truly integrated management system. We will also discuss some commonly-used risk assessment tools that can be applied to better understand any potential process risk or known process failure.

MANAGEMENT

The Manager's Role

Too many managers need to manage the delegation of their responsibilities more carefully. There is a fine line between delegation and abandonment of responsibilities. Management via committee is often a way to avoid making decisions or taking responsibility. Managers must lead, establish policy, take responsibility, and remove barriers that impede the safe and efficient execution of the work process.

How do you manage process design, enhance it, and add value to the process? Management must maintain the facility and train supervisors and employees to control the work system? The critical performance indicators must be identified and measured as part of the risk management process.

The Supervisor's Role

The supervisor is a key player, and some problems are beyond his or her control. However, a majority of problems in safety, quality, and production can be resolved here.

Supervisory Checklists. Use critical job performance checklists as an observation reminder of what to look for and evaluate, for example:

- Use of PPE
- Orderliness
- Use of Tools/equipment
- Actions and positions of people, as required by Standard Operating Procedures (SOP)
- Procedural compliance

Supervisory Control. Accidents usually result from both substandard conditions and acts. Supervisors must perform critical task observations, and take immediate corrective actions, if something does not conform. It's equally important to recognize and reinforce positive behavior and conditions. In addition, managers and supervisors must understand that oftentimes the multiple root cause of nonconformance and accidents is inadequate or poorly designed process inputs, as illustrated in Figure 1. These conflicting demands within the system or process generate mixed communications and force supervisors and employees into a position of substandard performance.

Effective Programs

Many programs are designed to reward or recognize employees who don't have injuries. More successful and permanent programs focus on the things people do (or don't do) to cause injuries and on the process inputs that are antecedents to job performance/behaviors. Positive reinforcement of desired behaviors as soon as possible after they have occurred is the most powerful way to ensure sustained process performance.

Communications and Training

- Share lessons learned from accident or incident investigations
- Stand-up or tailgate safety meetings
- Meetings should be regularly scheduled and have a written agenda
- Senior management involvement and work center employees involvement are both critical
- Record keeping/documentation
- Well-designed procedures (SOPs/PCIs) and training are critical

Quarterly mini-audits are necessary to evaluate program effectiveness and reduce risking facility management forgetting what the goals are, or their roles in preventing these potential losses. Involvement of employees is critical if the process is to be successful. After the first year you should begin to see the program start to become part of the company's culture.

Documentation and record keeping are essential for preventing unnecessary losses when false claims are made against the company. Records can also become the bases for process failure investigations and sharing the "lessons learned". Therefore, they become a tool for the development of more effective procedures. The quality and EH&S perspective of the work process discussion in this book is only an illustration of the issues and common elements of process management. I want to challenge managers who are struggling with quality and the increasing costs of injuries or environmental problems to consider these points:

- What can be done to shift the existing process paradigms?
- How can the work process be simplified?
- How can we improve the work processes, reduce redundancy, and identify and eliminate EH&S risks?

In addition, managers need to keep in mind the total costs (direct and indirect) of incidents or process nonconformances, the direct impact this has to the "bottom line," and the volume of sales/revenues necessary to make-up for these losses.

Downtime, low productivity, poor quality, injuries, environmental risk, damage to equipment or facilities, and product warranty or liability losses can all be controlled with a well-designed and implemented "quality and EH&S management system."

Management directly controls about 80% of OH&S risks. An effective program initiated with an audit of the workplace can identify the basic causes of potential process failures at the process-input stage as illustrated in Figure 1. It is these key elements that create the work environment that ultimately produces a product or service in conformance with requirements or creates an environment that encourages substandard performance. It is really a management choice.

Close management oversight at the output end of the process is also needed to ensure product quality and to control the risk of noncompliance with EH&S requirements. Reinforcement can be achieved by use of frequent job observations using supervisors and employees to observe critical job-tasks for conformance to requirements. Supervisors should provide immediate feedback and reinforcement of good performance or take corrective/preventive actions and provide process "coaching," as needed for observed non-performance.

As illustrated in Figure 7, the direct costs of accidents or improper handling of hazardous materials on the job are only the tip of the iceberg and can represent only 10 to 25% percent of actual losses. Like the iceberg, hidden costs of accidents are not visible on the surface but are there just the same. Crosby and other quality experts have estimated that the cost of process nonconformance can be as high as 20–40% for manufacturing industries and 50–70% for service industries.

Today it is becoming increasingly evident that neither the business nor the personal liability of the chief executive officer can be profitably transferred. In some cases transfer is not possible at any cost; for example, if the Environmental Protection Agency (EPA) or local law enforcement activities include criminal charges. Tort or criminal actions result when there is evidence of management's failure to exercise reasonable standards of care or a failure to do something when

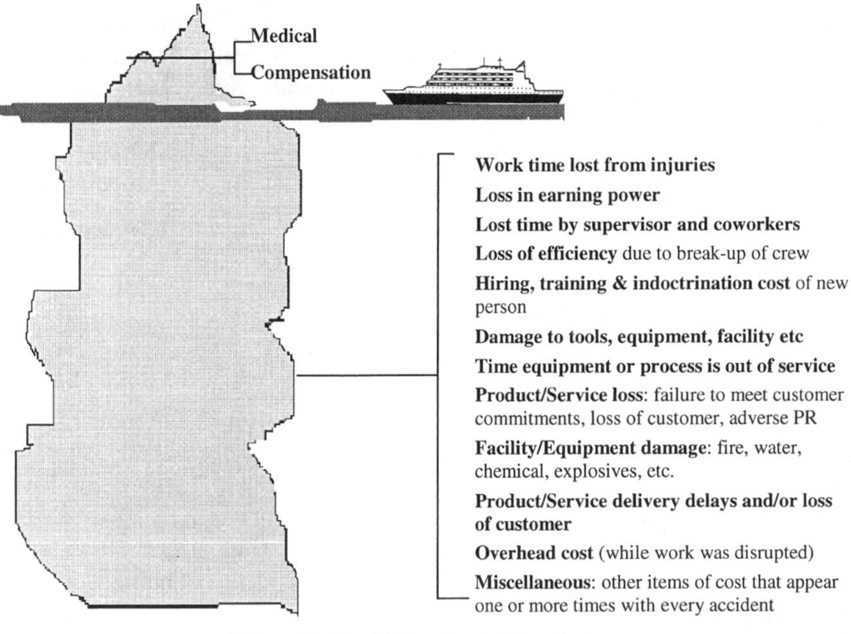

Figure 7 *The Hidden Cost of Accidents.*

there is a duty to act, which in turn gives rise to criminal charges or damage suits (in spite of the exclusive remedy provision of workers' compensation).

Multi-national companies will find it impossible to build a defense for negligence or criminal liability by arguing that they were not aware of requirements when it can be clearly demonstrated that their facilities in the United States or other countries are in compliance. A good example is the use of NFPA 101 "Life Safety" standards for buildings in the United States and the failure to meet these "Best Practice" guidelines in other countries.

Criminal liability is a realistic concern for senior managers and has been demonstrated in an increasing number of criminal actions. Executives have even been convicted of murder and sentenced to 20 years in prison for their failure to initiate controls and warn employees of chemical hazards. A Chicago district attorney sought indictments and convicted these managers because of their failure to warn Polish immigrants of hazardous materials that caused a worker's death. Other cases in Chicago have resulted in counts of reckless conduct and aggravated battery against an electrical works manufacturer for the acid burns of 125 employees. These facts underscore the importance of providing a "healthy" workplace and developing a structured OH&S program. More recently murder charges were filed against the corporation responsible for the oxygen canister causing the Value Jet crash.

During the past 15 years, we've seen district attorneys increasingly resort to criminal actions when there has been an occupational fatality or serious injury. Law enforcement officials in Los Angeles take industrial fatalities so seriously that they have formed a special prosecution unit. The Los Angeles county unit has a stern message for employers regarding failures to comply with EH&S laws, which is to comply or legal actions will be initiated. These prosecutors are seeking criminal convictions, large penalties, and prison sentences. They believe that employers should be liable for their actions, and that they must raise their standard of care.

In September 1990, California implemented two assembly bills, which were tagged as "Be a manager, go to jail." These laws were added to California's penal and labor codes. AB 2249 makes it a crime to conceal serious hazards and states, "That a manager ... is guilty of a misdemeanor or felony, if the corporation or manager has actual knowledge of a serious concealed danger that is subject to the regulatory authority of an appropriate agency and is associated with that product or a component of that product or business practice and knowingly fails to inform the Division of Occupational Safety and Health and warn its affected employees."

ESTIMATING INDIRECT COST

Protecting employee safety must always be a primary objective followed by regulatory compliance. However, effective execution of IMS requires the EH&S professional to more diligently communicate risks, known losses, and proposed control alternatives in financial terms to enhance the chances of budgetary approval in an environment where there is competition for resources. Most companies do not adequately document both the direct and indirect costs of accidents or process nonconformance (total cost). They therefore tend to understate the impact of these losses and fail to hold the work center accountable for the losses, absorbing the costs in a corporate or facility "risk transfer" that is not recognized locally.

Often accurate documentation of all (direct and indirect) costs is not readily available. Therefore, reasonable, indirect cost estimates, and the assumptions used to define them, may be needed. You should temporarily track indirect costs while developing a loss history of direct and indirect costs by accident type. For ease of reporting use a ratio of indirect costs to establish meaningful "total cost" estimates. It is also important to create supervisor accountability by measuring costs down to the facility or work center, not unlike other operational budget items.

As a rule of thumb, companies can assume that indirect costs will be 4–10 times the documented direct costs of process nonconformance and accidents. Managers can develop accident or incident investigation forms and procedures to more accurately document these costs or scrutinize past non-conformances to determine relatively accurate estimates of indirect costs associated with different types of process failures. Once the total costs of process nonconformance or accidents are known, the cost should be charged back to the work center responsible for the incident. Figure 8 illustrates the significant impact a relatively small direct cost can have to a company and its profits and cost of sales.

Management must understand the financial impact of accidents or process nonconformance all the way down to the work center supervisor. Other "benchmarks" that could define loss rates and the total costs of injuries are

Loss Rates — Cost of Injuries

- Dollars lost per 200,000 work hours
- Dollars lost per unit of production, for example, total accident cost divided by the number of packages shipped or revenues produced during this period
- Cost as a percentage of profits

Both employees and employers are become increasingly concerned with the long-range effects of some materials and products used in the workplace.

Example: At 10% Profit Margin
$15,000 loss (direct cost)
4:1 ratio of indirect to direct costs, causes $60,000 indirect costs
A facility will need $750,000 in additional sales to make up this loss!

Figure 8 *Impact to Profits.*

Additional concerns include hazardous materials and waste, especially carcinogens, mutagens, or teratogens. Human exposure to new chemicals has grown from an annual production of 7.3 billion pounds of synthetic chemicals in 1940 to more than 350 billion pounds today. About 500 chemicals are regulated by the OSHA) and many others fall under the regulatory scrutiny of the EPA. There are 14,000 to 15,000 toxic substances commonly used at work sites, whereas hundreds of new ones are added each year.

- In addition, of the approximately 50,000 commercially available chemicals or chemical products, 80% have little or no reliable toxicological information. According to the American Industrial Hygiene Association, a new toxic agent is introduced into the workplace every 20 minutes. It is estimated that 125,000 workers will suffer from occupational diseases every year. Due to the long lag times between exposure and onset of symptoms, it will be the year 2025 before we realize the full impact of substances introduced into the work environment today. The list below is considered to be the most serious occupational disease and injury categories
- Occupational lung diseases, including lung cancer, pneumoconiosis, and asthma
- Cumulative trauma disorders
- Traumatic deaths, amputations, fracture, and eye losses
- Occupational cancer other than lung cancer
- Cardiovascular disease, including myocardial infarction, stroke, and hypertension
- Reproductive illness
- Neurotoxic illness
- Noise-induced hearing loss
- Dermatological problems, including dermatitis, burns, contusions, and lacerations
- Psychological disorders

Employers in industry today need to concentrate on three areas in order to manage workplace risks effectively. First, they need to define the potential risks and need for controls in facilities, equipment, and material processes. Second, employers must educate supervisors and employees on how to recognize risks and control them. And finally, they must evaluate hazardous material/waste storage, transportation, and disposal risks, including safety hazard abatement and environmental re-mediation plans. They must be able to evaluate their company's culture and primary process risk factors. The culture and environment is created in the workplace in a way that either supports great performance or encourages substandard performance by ineffective provision of the process inputs (illustrated in Figure 1) or by poor or mixed communications.

SAFETY IS PRODUCTIVITY

An effective OH&S program is good management and should be viewed as part of the employer's overall quality assurance program. Real productivity is the balance between striving for excellence and protecting human resources. The chairman of the board, president, and general manger of the company or facility must give priority to environmental health and safety control programs. Managers must take this responsibility as seriously as they do meeting schedules, producing superior products or controlling their costs in other areas.

A good OH&S program first requires a close review of overhead expenses for compensation, environmental differential pay, health care benefits, and employee assistance programs to determine:

- Major employee/corporation risk factors
- Necessary controls to protect the employee resources
- Engineering controls
- Administrative controls
- Personal protective equipment
- Supervisor/employee training
- Wellness/fitness training
- Family/employee assistance programs
- Incentive programs to control medical benefit expenses
- Medical case management and compensation claims management

U.S. OSHA health standards also require that employers monitor potential exposures, which means collecting air samples, making bulk material analyses, and taking noise measurements. If exposures exceed the permissible exposure limit (PEL), employers must

- Establish engineering controls to reduce exposures, substitute the material for less toxic hazardous substances, or change the process
- Establish a personal protective equipment or respirator program
- Establish a medical surveillance program
- Establish and maintain a record-keeping system for up to 40 years
- Periodically monitor at intervals of 30 days to 6 months as prescribed in each specific standard (This process is usually required until two consecutive rounds of monitoring data indicate that potential exposures have been controlled to a level 50% below PEL)

Employers should make sure they have documentation to prove that potential safety and health hazards have been identified and controlled at a level below the acceptable threshold limit values. Record keeping is essential because the employer must be able to demonstrate that potential hazards have been evaluated and that employees understand and have received training on the nature of the hazard, process requirements, and controls.

A reasonable effort must be made to ensure that workers comprehend the relevant hazard/risks associated with their jobs or tasks and control information. Companies need to devote appropriate time and resources to gathering, interpreting, and explaining chemical workplace hazards to employees and community emergency response agencies or coordinators. Employers and employees must have enough information to make intelligent decisions. The EPA's Community Right-To-Know policy also requires that companies provide their community with emergency planning organization and information about the chemicals used or stored on their property.

CONTROLLING RISK AND COST

If the regulations aren't serious enough to motivate actions, litigation will be. Cumulative injury claims have been increasing faster than any other claim category. Even though workers' compensation was designed not only to provide benefits to injured employees but also to limit the liability of the employer, courts have found many loopholes in the concept of "exclusive remedy." As an example, the Ohio courts decided in *Blakenship* vs. *Cincinnati Milacron Chemicals* that workers' compensation does not extend immunity from civil suits when the employee alleges intentional misconduct or failure to act. In this case, eight employees who contracted occupational diseases argued that the company was aware of the chemical hazard and failed to take corrective actions or warn the workers.

Proactively, employers can control the costs associated with employee health benefit programs and compensation by establishing comprehensive health and safety programs, including "wellness" training and employee assistance programs. These help increase productivity, while capping or reducing the costs of compensation or health benefit coverage, by keeping an employee safe, healthy, and in control of his or her lifestyle.

Many managers look at OH&S as a difficult and possibly an uncontrollable problem or, at a minimum, "expensive to resolve." The most difficult part of the problem is confronting and understanding these issues and the specific workplace processes. Once you have defined the process problem, the controls will become apparent, allowing you to treat safety concerns no differently than any other management or production risk.

Management also must learn to follow up on all cases of workplace injury to learn the multiple root causes of the accident or process non-conformance to prevent a re-occurrence in the future. Supervisors cannot divorce themselves from injured employees; they should show them concern and accommodate disabilities so that a return to work is possible.

Benefits that the employer can expect from a well-managed OH&S program:

- Reduction in medical/compensation costs or employer tort action
- Reduction in injury/illness frequency and severity rates

- Reduction in sick-leave use and other absenteeism
- Reduction in environmental risk liability
- Improved safety and health programs
- Improved employee morale and community image
- Reduction in training costs
- Improved employee retention
- Improved quality, less rework
- Reduced risk of regulatory civil or criminal actions

To achieve these benefits and maximize the quality of an organization, employers must be prepared to do several things. First, health, safety, and environmental risks should be identified and documented, and plans must be made for controls in the workplace. Environmental health and safety should be intricately linked to the company's quality improvement process (QIP), if one exists. If a QIP does not exist, starting one would be beneficial, because the additional management attention required would have a synergistic effect in reducing losses and increasing profits. In-depth management audits should then be conducted annually. Integration of QEH&S is recommended.

Also, the employer should comply with OSHA, EPA, and Department of Transportation laws, as well as reduce compensation liability. If an injury occurs, the employer should try to assist in the return of the employee to his or her full potential. In any event, employers can enhance the productivity of the organization by increasing the general health, well-being, and fitness of managers and employees.

The most important thing is to face the problem — or potential problem — and to treat it simply as part of the overall management responsibility and quality assurance program. The benefits are apparent and can be realized once the mystery or misunderstanding is clarified.

Protecting shareholder value and the health and safety of our staff and customers who use our products or come to our facilities is critical to maintaining corporate health. The company that places short-term gains ahead of these principles will ultimately see its business fail. Risk management and process control must be linked to operational decision making. Process and risk management cross-corporate functional lines. In addition product "life cycle" risk assessments, including cost estimates, should be performed for all new products or existing product lines (cradle-to-grave HM/HW management).

To be successful the operations mangers must see the risk manager or process control manager as a member of their team. This perspective is necessary if we want to leverage and integrate the technical and managerial skills that are committed to disciplines such as quality management, EH&S, and risk management. The total cost of risk includes "risk transfer" premiums, administrative costs, and retained losses that are ultimately absorbed directly or indirectly by the business unit or work center as a direct cost or increase to overhead. We must be able to learn from the past, recognize, and manage risks in existing or anticipated processes or operations to be successful.

OH&S Management Systems Requirements. Setting policy and process standards to manage operational risk reduction and quality enhancement, including regulatory compliance. However, process must have a performance versus regulatory focus. Well-managed systems generate compliance, whereas management focus on regulatory compliance tends to generate incomplete and costly compliance and ineffectively contributing to operational efficiency. Risk identification and assessment is necessary to ensure that a prioritized plan is established to improve a facilities process performance. No company can successfully try to correct all performance issues at the same time.

Clear assignment of responsibility and accountability is necessary to support the integration of quality and EH&S Management Systems, including:

- Communications
- Training
- Performance measurement system (What gets measured gets done, and usually with improved quality)
- Audits and job observation are critical
- Document control and record keeping
- Conformance evaluation, management review, and feedback with routine follow-up of assigned actions until completed
- Contingency-emergency preparation

The process of quality and environmental health improvement starts with an audit (gap analysis) of the facility and baseline cultural survey. The purpose of the audit is to understand the company's risk, the cost of their losses, and how the existing management system deals with these potential problems. Quality and process management training (including environmental health and safety) should be conducted at all organizational levels. Process management enhancement action teams will be needed at critical work centers.

Your goal will be to develop a management system that decreases workers' compensation claims, health care costs, rework, and the volume of waste generated—particularly hazardous waste. To effectively enhance productivity and reduce rework or lost time due to injuries, all levels of management should be involved, including senior management. Frequent and supportive meetings will significantly empower these process enhancement teams to create the results each facility will want to achieve. Senior management must be visibly active and committed to the process if it is expected to succeed.

Everyone must understand that all work is a process and that we all have internal and external customers with specific needs. If the process is failing as reflected in increased rework or accident incident rates, it needs to be defined and reengineered. In addition we must look at the work environment to determine what critical process inputs are causing nonconformances or incidents.

We accomplish this objective by using resources such as process enhancement teams or employee-facilitated risk assessments, systems safety reviews or job task analysis, process flow diagrams, critical job observation, critiques, root cause or multiple root cause analysis or process nonconformance and accident/incident investigation procedures.

In addition supervisors must be empowered and must understand what controls they have over the work process. Senior management involvement must be demonstrated by participation in periodic inspections of the facility and work processes — risk assessments with the tools identified above.

PETs can use tools such as brainstorming — for example, facility risk assessments, Parieto, or actual loss data — the OSHA 200 log categorizes and prioritizes risks and related process improvement efforts, which include inspections and detailed process task observations. Focus initially on high-risk operations. Create a positive impact and savings ASAP!!! Start with those processes or work areas that represent the greatest risk (don't try to control or improve everything at once). Prioritize your risk and the available resources to be applied to mitigating the risk.

To advance, management systems must require that companies' establish or document processes to review/investigate process failure, rework, and/or incident/accidents. These review procedures should include process nonconformance or incident multiple root cause analysis, systems readiness, cost of non-conformance, and system for ensuring that appropriate corrective actions are implemented and followed-up, including sharing of lessons learned with "stakeholders." In addition, effective performance measurement tools and management system are necessary.

- Process charting of nonconformances or accidents/incidents
- Job critical control or behavioral safety observations
- OSH programs, such as respiratory protection and manuals
- Business plans for waste reduction or minimization
- Accident injury prevention programs
- Supervisor_coaching
- Assisting supervisors to develop and use critical work center process observation check sheets
- Facilitating the continuous review and improvement of critical processes to eliminate rework, improve quality, reduce accidents
- Conducting periodic program audits
- Documenting corrective actions or process improvement
- Assisting managers to ensure corrective actions or process changes are effectively implemented

The bottom line is that the smart process or quality management system cannot treat safety as a separate issue. Therefore, effective programs will integrate quality and EH&S (QEH&S).

ACTION PLAN CONSIDERATIONS

- Do you have qualified EH&S personnel on your staff, or have you identified consultants that you are comfortable with and understand management systems and management audits versus regulatory audits?
- How are jobs planned, and are EH&S concerns an integral part of the job or product planning/design processes?
- Are employees and supervisors trained and equipped with the tools and resources to do their jobs safely, and without causing environmental liabilities or rework?
- Have you identified process EH&S risks, and have interim and long-term corrective/preventive actions been implemented?
- Do you know the direct and indirect costs of process nonconformance and accidents, and are these costs charged back to work centers?
- How are chemicals managed from the design, procurement, and receipt process to the ultimate use in production, including control of production waste streams?

3

Management Commitment: Policy and Planning

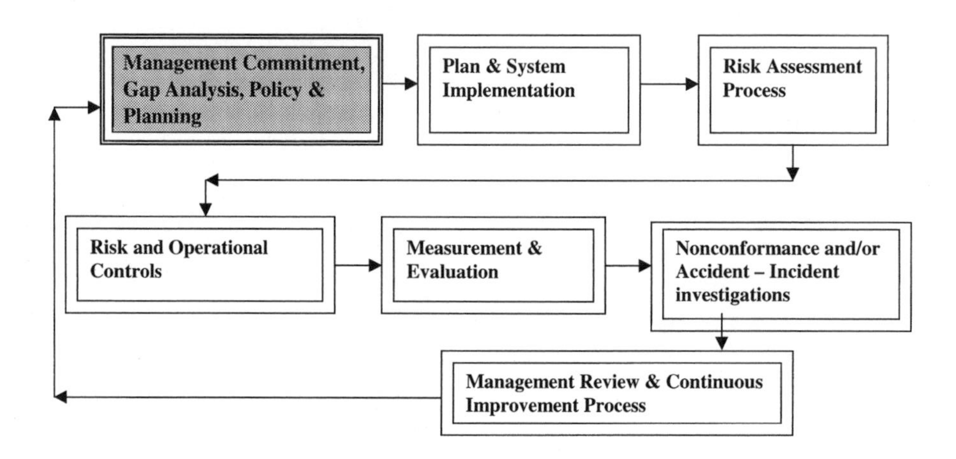

> The industry is in continuous development, and so are the tempers of consumers. Both demand more and better quality...
>
> —*Egyptian Cotton Exporting Co.*

Organizations need focus and a clearly communicated direction from their leaders. If you want superior quality it must be designed into the process. The level of success any company will experience in improving quality or environmental health and safety (EH&S) will be in direct proportion to the level of commitment and involvement of its senior managers to the process. Policy alone without the active involvement and leadership of managers will ensure disappointing results, which is probably one of the reasons so many management systems fail. Gillette's leadership builds safety into the work process by using processes such as "Safety Through Design," which has six phases to design nonconformance and incident prevention into the new process.

- Planning, which includes risk assessments and participation from the product design engineer, EH&S, quality, and work center or process employees and supervisors
- Functional specifications
- Design
- Construction and installation
- Testing or functional debugging
- Production acceptance

Planning is secondary to risk assessments, audits, and gap analysis. These tools provide management with effective diagnostics to identify systems gaps or process risks. Risk assessments will provide work centers with ranked risks and a prioritized action plan to incorporate into their overall facility planning and management review process. A more detailed discussion of audits is discussed in Chapter 7, and risk assessments are covered in Chapter 5. Employee involvement in the performance enhancement team (PET) or other facilitated group risk assessment process is critical. Employees should also be involved in accident investigations (discussed in Chapter 8), critical job or task observations (see Chapter 7), and product "life cycle" risk evaluations.

All management systems require senior management to step to the helm. This chapter discusses aspects of management commitment and the establishment of policy. For the most part we recommend a process that is very similar to the requirements addressed in ISO 9000/14000 and by quality leaders such as Deming, Juran, and Crosby.

"Taming the Dragon" requires management commitment to establish and execute consistent policy for process quality and EH&S improvement. Top-down management is critical, as the dragon's heart, the first-line managers or supervisors, will need support to make the critical process improvements they know are necessary. Support will be necessary to change processes, replace tools, ensure maintenance, and possibly redesign facilities and improve training, procedures, and communications.

Without management's support first-line managers will become frustrated and will abandon the process or at best passively support it. Process improvement programs cannot simply focus on employee behavior and empowerment. Successful programs will require a focus on both the management and supervision levels. These two organizational levels control the organization's resources, including employees and all of the facility tools, equipment, procedures, and training to perform work. Management also controls the company's culture and ultimately the behavior or performance of the employees both in quality and

Figure 9 *Orchestras do not lead or direct themselves and neither do organizations.*

EH&S improvement. Most importantly, management must be sensitive to establishing incompatible goals.

The commitment and direction from management should come in the form of policy and budgetary support, including the assignment of key personnel. Policy should be consistent with the organization's operational, QEH&S risks. Management must also establish a system to continuously evaluate its performance and to understand the impact of its activities on the environment, employees, community, and customers.

Leadership must identify and clearly define core business processes, critical procedures, points of risk or potential deviation, and processes for improving them. This approach must be done on prioritized bases. No one ever succeeds by trying to "fix" everything at once. Therefore the risk assessment and prioritization process discussed in Chapter 5 will be a critical tool for Process Enhancement Teams PETs. If prioritization of the risk assessment/control process — including establishment of goals and objectives — is not integrated, separate quality EH&S process enhancement teams may in their respective vacuums work against each other. It is therefore important to rank corrective and preventive actions and align related goals and process improvement efforts.

As discussed earlier, the Chinese believe that it is better to take 1,000 small steps than to leap into the abyss. Management's vision statement and commitment to process improvement must be translated into executable long-term goals, "SMART" (specific, measurable, achievable, and agreed-to strategies, including target dates) short-term objectives. Typical goals might be:

- Increase market share and revenues.
- Reduce rework and other process non-conformances such as environmental health and safety losses.
- Reduce waste, injuries, and uncontrolled environmental releases, recycle materials.
- Reduce the design and cycle time for delivering new innovative customer-driven products and services.

The continuous improvement process must also consider regulatory requirements as well as the prevention of uncontrolled releases of hazardous materials-hazardous waste (HM/HW). The management system must document procedures to monitor performance and ensure that potential exposures to HM/HW are controlled, until the potential risks are eliminated.

ISO 9001 and 140001 will require documentation, such as manual and written procedures defining the requirements and continuous improvement process for an integrated QEH&S enhancement process. The QEH&S manual should establish policy, incorporate or reference work procedures and general QEH&S requirements. The manual should include all relevant programs that support the QEH&S management systems and must be available to operations essential to the effective functioning of the QEH&S management systems.

Companies that will succeed in the future, maintaining or advancing their current market and financial position in our global community, will be those that see QEH&S as a condition for economic growth. "Total quality environmental management (TQEM) may result in increased profitability and increased environmental responsibility, including a general improvement in production quality, production efficiency, and environmental standards... Companies are recognizing that by focusing on process and product design, including QEH&S rather than on pollution control and cleanup, they increase profitability." (Epstein, 1998)

QEH&S planning must produce proactive rather than reactive companies focusing on planning and product life cycle risk rather than regulatory compliance. Product/service or life cycle design and planning must focus on eliminating risk, waste, rework, and EH&S liabilities. Focus must be on making decisions that will improve the profitability and sustainability of the company's products and business, while minimizing risk through a process of "continuous improvement." "Regulatory-driven waste reduction programs cost three times more than voluntary driven ones for the same benefit." (Ref. 20) QEH&S and product-service design management must focus on:

- Providing products and services that meet or exceed customer expectations.
- Reducing or eliminating EH&S risk to employees, communities, and consumers.
- Reusing materials to minimize the generation of waste and its liabilities and product cost.
- Recycling to the extent possible the waste that cannot be reused.

"It is generally desirable for the senior EH&S (QEH&S) officer to have direct access to both the board of directors, and the CEO, and not be in a legal function. Often the legal compliance approach is detrimental to the planning necessary to move towards environmental sensitivity in corporate decision making. It has become clear that numerous benefits do accrue to companies that make capital, process, and product improvements as a result of the analysis of corporate environmental impacts. These companies usually go beyond regulatory compliance and make voluntary improvements" (Epstein, 1998).

COMMITMENT AND POLICY

An organization should define its policy and ensure commitment to its management system. Execution of management's policy should be evident in internal documents and procedures, including management, supervisors, and staff training. A member of the executive management team should have documented overall responsibility for ensuring the effective execution of the QEH&S enhancement process. This responsibility must include maintenance and distribution of information regarding the continuous improvement process. Management system auditors will want to verify the documentation of process records and must interview employees to evaluate their understanding of required procedures and policy as well as their knowledge of and participation in the process.

To be successful with the QEH&S process enhancement, management at all levels must demonstrate their commitment to the system by demonstrating the following.

- Active, visible participation in the process.
- Providing adequate resources to the management system.
- Integrity — following through on commitments.
- Setting an example of how quality and EH&S standards will be executed and maintained (walking the talk).

Drafters of any management system policy must ensure that the policy reflects their beliefs and the focus they want to generate in the operation. To be successful, the policy and all supporting procedures, goals, and objectives must be relevant and have as much concurrence as possible from employees and other "stakeholders," especially internal and external customers. In addition it must establish a commitment to understand and comply with all company standards, including a process for conforming to applicable legislation, risk reduction, and general process improvement policies.

The community may also need to participate in the development of policy regarding emergency response, or the release of HM/HW. In manufacturing facilities, effective community response can be severely compromised if the local emergency response departments do not have the necessary information to respond to chemical releases or fires on a property containing HM/HW. Personnel responsible for taking control in an emergency must be trained and known in the organization. In addition the staff must have clearly defined response and communication plans in the event of an emergency.

In some situations lack of communication could turn an emergency situation into a disaster with serious injuries and fatalities. Therefore, the policy should include a commitment to identifying and communicating to all employees, contractors, and the general public the risks associated with work activities and the controls in place or required to mitigate the risk.

The management system policy and all supporting documents or procedures should have a process for review and continual improvement. Management can accomplish this objective by establishing, audits, and management reviews of QEH&S performance with regards to their effectiveness in executing objectives/targets, and conformance with legislation or procedural changes. The management system policy should reference responsibilities of management, supervisors, employees, contractors and suppliers for the QEH&S improvement process.

COMMUNICATION

Communication, especially employee communications, is critical. Sharing lessons learned from process nonconformance investigations, risk assessments, critical job or task observations, process instructions, standard operating procedures (SOPs), training, and supervision all provide significant opportunities for supervisor and employee communications. Internal reporting and investigation procedures (see Chapter 8) should be established to cover these aspects:

- Incident occurrence reporting and investigation.
- Rework or customer complaint reports.
- Non-conformance reporting.
- Health and safety performance reporting.
- Hazard identification reporting, including defining the multiple root causes of the potential hazard or documented process non-conformance.
- Reporting of environmental releases of HM/HW or process control nonconformances.
- Risk assessments and risk prioritization.
- Process changes.
- Sharing of lessons learned.

There should be procedures for ensuring that relevant QEH&S information from the identification of the multiple root causes of the process non-conformance is communicated to all involved staff, management, and contractors. "Sharing lessons learned" as a result of a risk assessment and accident or process nonconformance investigation are critical. Management must demonstrate a commitment and a process for it, as well as a means of communicating changes to process design, materials, or procedures. The communications process must be designed to ensure that critical facility, regulatory, or process information, including "lessons learned," are effectively relayed to people inside and outside the organization who require it.

Other necessary communications could include the following:

- Posting the results of management system audits, monitoring, and management reviews, especially for those that are responsible for or have a stake in the organization's performance.
- Identifying and receiving relevant QEH&S information from outside the organization, such as changes to regulations.

RESPONSIBILITY AND AUTHORITY TO ACT

Senior management must establish a management system to ensure delivery of quality products and services and to control the potential for process losses, including HM/HW releases and injuries to employees, contractors, or customers. The policy should include process-production goals for achieving QEH&S objectives and milestones and for involving employees in the process.

The QEH&S manual must define responsibilities, delegate authority to act, and clearly define the reporting relationships of all personnel, including responsibilities imposed by legislation or regulations. Most manufacturing operations will require either internal expertise in QEH&S improvement or a commitment to have access to outside/consulting expertise.

To be effective, process improvement procedures must support supervisors in meeting their schedule and budgetary requirements and should clearly define the responsibility, authority, and relationships of personnel who manage, identify, and evaluate process control, including supervisors, contractors, and vendors. The QEH&S and/or risk managers must have defined responsibility and authority for establishing, implementing, and maintaining the requirements of this management system, and they should be actively working together to evaluate and improve work processes, while eliminating potential losses. The performances of the operations, quality, and EH&S managers and all supervisors should be regularly evaluated to verify that they are achieving clearly defined process improvement/risk reduction goals and objectives.

The management system should also include a process for amendments due to regulatory or process changes, or any other significant modification in activity, product, or service delivery requirements. The operations, quality, and EH&S manager's responsibilities and authority, including interdepartmental roles, must be clearly defined.

The role and accountability of management and supervisory personnel to the QEH&S process must be clearly communicated to ensure effective process improvement/quality and EH&S performance. QEH&S performance should be included in the organization's annual report or equivalent "stakeholder" communications documents or reports.

Senior management should demonstrate its leadership and commitment to the management system by allocating adequate resources, including budget, and by defining responsibilities and the authority for ensuring the execution of the system and process conformance. Responsibility and resource assignment should include authority and the necessary tools, finances, equipment, staff, and training to effectively implement process improvement requirements and policy to ensure the execution of a continuous improvement system.

SUMMARY OF KEY QEH&S MANAGEMENT SYSTEM POLICY REQUIREMENTS

The following points can aid in establishing or rewriting an organization's policy.

- Policy must be integral to the organization's overall mission statement, vision, core values and beliefs, management activities, products, and services. Managers and supervisors must be accountable for successful execution of the "process improvement" system.
- The management system policy must address accountability in terms of allocating resources; delegating, delivering, and reviewing commitments; and setting objectives and performance targets to minimize process risk while enhancing quality and process performance, including EH&S.
- The policy should also include a statement of compliance or due diligence with regards to relevant EHS legislation, regulation, corporate or industry guideline, and requirements.
- Planning must consider all "stakeholders" and include needs of special groups, including the community and various stakeholders; shift workers; job rotations; new projects; and product and service delivery.

Planning moves management from good intentions or wishes to the greater challenge of reality and improving the existing position of the organization, including QEH&S. Effective planning must include the establishment of specific, measurable, achievable, and reasonable target dates (SMART) goals and objectives. Management owns the process and all of the resources applied. Therefore management must be committed to the planning and system implementation, as decisions will be needed regarding the timing and allocation of resources.

MANAGEMENT JOB INFLUENCES

Management controls the workplace and all of the processes and resources applied to completing the job. Within this environment, employees are subjected to conditions and influences that will promote and support doing the job safely and "right the first time," or it will influence substandard performance. If management cannot purchase the "right" material, tools, and equipment or provide needed maintenance, the job will be severely compromised before it starts. Communications, including training, procedures, and supervision, must then be consistently applied for employees to successfully perform their jobs safely and meet customer expectations. Management must therefore commit to the following.

- Providing the right tools, equipment, and materials, including maintenance.
- Providing better reference documents, training, publications, and guidelines.
- Improving job or task procedures.
- Redesigning equipment or processes to enhance efficiency and quality while reducing EH&S risk.

Organizations should identify competent management and staff for developing plans to fulfill its QEH&S policy, goals, and objectives, including assessment of risks and their controls. This plan should include review of facility, tool, and equipment readiness or applicability. It should also include the review or development of well-defined process control instructions for "at-risk" operations. The successful implementation and operation of management systems requires an effective management planning process. Systems should also be evaluated to ensure that the correct procedures, tools, and equipment are available and that work centers have effective maintenance, training, and supervision.

In addition to the initial planning phase, procedures should be established to ensure that planning is undertaken in the ongoing operations, including periodic management review (e.g., annual) of the organization. Where there are changes to the activities, products, or services of the organization (e.g., introduction of new products) or significant changes in operating conditions such as a change of location, facilities will need procedures ensuring that risk assessments are done to evaluate the impact of the process change. This plan could include process change analysis, a system safety or readiness review to confirm the adequacy of procedures, facility, tools and equipment, maintenance, training, communications, and management (supervision) to perform the work processes.

Employees must be made aware of possible QEH&S risks in the work environment, including product or service delivery risks and physical and chemical hazards. They should understand these risks as they relate to their work and recognize and take action to prevent work practices or activities likely to lead to accidents or process nonconformance.

Planning and Implementation: What, Who, How, and When

To be effective, management must consider a process for continual improvement and change. As I indicated earlier, it is not wise to try to change the world overnight. Therefore, most process changes will be much more effective if they result from a prioritized ranking of process or facility operational risks. In addition, objectives must define the category of objective and the measurable

TABLE 7. Process Management SMART Planning

Type of Objective	Specific Measurable and Time-Bound Requirements (Examples)	Measurement (Examples)	Time Period or Target Date
Quality, Accuracy or Product-Service or Process Reliability	Establish a system to measure product performance against customer requirements, including cost of nonconformance, ensure that process defects do not exceed statistical control level, all nonconformances will be evaluated to determine the multiple root causes and to initiate cost-effective controls	Number of Nonformances Severity of Nonconformance or Percentage of nonconformances, number of complaints or process failures	3/1/00
Production or Service Time	Ensure that service design and deliver time are reduced by 15% in the Manhattan division	Number and percentage of commitments achieved, service or delivery time	6/15/00
Cost Per Unit of Production	Establish a system to evaluate the financial impact of all process nonconformances and the cost per unit of production or percentage of PTOI	Cost of nonconfor-mances per unit of production, cost of waste, lost time, injuries	4/30/00

goals that will be accomplished by a specific "target date." Managers should also try to ensure that they have concurrence from a majority of the affected "stakeholders" before implementing changes. Table 7 is simply one way to document established goals and objectives. Internal monitoring should be established to ensure that target dates have been met and that there is a process to evaluate the executed actions effectiveness.

OBJECTIVES AND TARGETS

Organizations should establish and maintain procedures for documenting QEH&S system activities, goals, and their results, including periodic review for adequacy of conformance/execution. As appropriate, you should also consider the present/future value of risk controls when comparing the costs of control alternatives under consideration to ensure that the most cost-effective changes are made and that a clear assessment and recommendation process is communicated to senior management for approval. The cost assessment should include using defined assumptions for internal rates of return, depreciation, risk reduction, and inflation rates.

Combining targets and performance indicators produce "SMART" objectives that are specific, measurable, achievable, realistic, and time-bound. Objectives state what is intended and they establish a timeframe for conformance. A "goal" may be defined as a broad statement of an end result to be attained to satisfy the QEH&S policy. An "objective" is a specific initiative that is measurable, quantifiable, time-bound, and necessary to satisfy a goal. For example, a QEH&S goal may be to minimize lost workdays resulting from occupational injury. The stated objective might be to reduce lost workdays in Department X by 30% and Department Y by 65% over the next 12 months (See Table 6).

To achieve its QEH&S objectives, an organization should involve its people as well as focus and align its systems, strategies, resources, and structure. The management system is designed to ensure that system requirements are established and that appropriate performance reports are submitted to top management for review and system improvement. In allocating resources, organizations should develop procedures to track the benefits as well as the costs of their activities, products, or services.

Regulatory review, Compliance in the Planning Process

Management must comply with regulations that relate to business. This case is especially true for some environmental permit requirements and for the processing or handling of HM/HW. Failure to comply can in some cases result in criminal actions, penalties, and distractions to the business, and are thus a threat.

However, as mentioned earlier, there is a risk associated with making regulatory compliance (as opposed to process conformance) the primary objective of the management system. By focusing on process conformance and continual improvement, many regulatory requirements will be met through a sound process management system. Whereas focusing on regulations is often expensive and may do little to improve process conformance. SMART planning that focuses on process enhancement must be the first priority.

Management System Planning Considerations

- Establish and communicate policy and procedures.
- Determine the need for capital investments and prioritize and conduct financial evaluation of investment alternatives.
- Contract with approved suppliers, vendors, and manufactures.
- Evaluate business-operational risks and conduct prioritized facilitated contingency planning.
- Develop QEH&S manuals and operating SOP's.
- Develop production or process control instructions and employee–supervisor training.
- HM/HW shipping, storage, use, reuse, recycling, and disposal.

- HW treatment and disposal guidelines.
- Packaging guidelines.
- Energy use-management.
- QEH&S audits.
- QEH&S data, documentation management (information systems needs analysis).
- Customer, shareholder, community stakeholder communications, including marketing public relations.
- Management development and supervisor training.
- QEH&S department and division or facility professionals for QEH&S or establishing outsourcing-consulting needs.
- Process-job observation and production measurements-evaluations.
- Product design and life cycle reviews.
- Technology transfer and "sharing of lessons learned."

ACTION PLAN CONSIDERATIONS

- Review the overall business case for an integrated management system (IMS), including the future needs of each element of the separate systems (safety, quality, and environmental), and the adequacy of the existing systems to meet those needs.
- Pilot IMS in a segment of the organization or at a volunteer facility.
- Develop a phased implementation plan with specific milestones.
- Identify the key skills and training needed for each segment of the IMS that must be retained in the new system, work with human resources to develop a well-planned training schedule that appropriately reaches every level of the organization and is tailored to the specific needs of each segment of the organization and/or staff-management level.
- Determine measurable criteria to monitor and review the effectiveness of the IMS.

Action Plan

- Draft a management policy integrating quality and EH&S.
- Establish a Process Enhancement Team PET with senior managers, including managers for risk management, QEH&S, employee representatives, and other "stakeholders" as appropriate.
- Establish measurable goals and objectives, and plan performance targets.
- Establish nonconformance/accident reporting and investigation procedures and train all supervisors in the process.

- Establish process nonconformance cost documentation, including work center charge-back system.
- Establish measurable process enhancement goals for all supervisors and managers.
- Ensure that the PET, risk, quality, and EH&S managers have defined plans, resources, and approved budgets to execute the QEH&S policy and plans, taking into consideration the results of prioritized risk assessments.
- Evaluate applicable regulations and ensure that the facility complies with key regulatory and permit requirements.
- Develop a plan of action and milestones, assigning target dates to responsible managers.
- Establish a follow-up and implementation evaluation/reporting process.

4

Plan Implementation

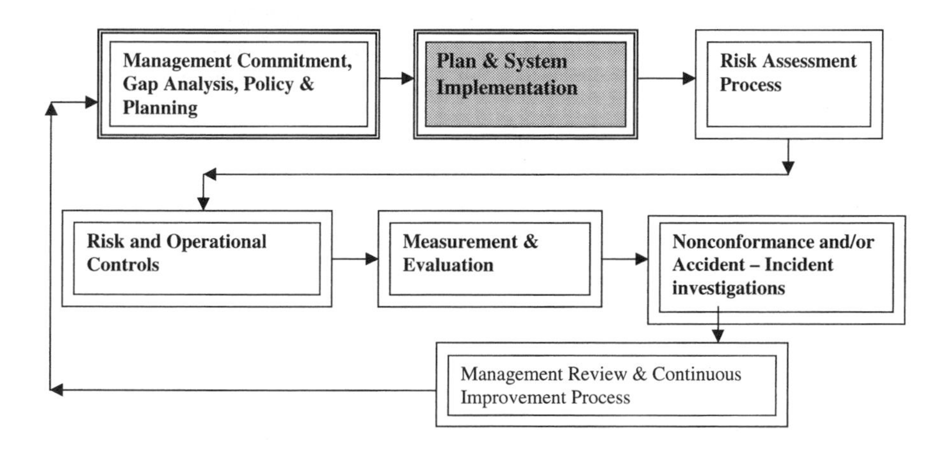

It's not enough to take steps which may some day lead to a goal, each step must be itself a goal and a step likewise.

—*Goethe*

The world owes all of its onward impulses to men ill at ease. The happy man inevitably confines himself within ancient limits.

—*Nathaniel Hawthorne*

This chapter focuses on action plan implementation. At this stage, you mill have conducted your Gap Analysis audits and culture reviews or an assessment of primary process risk factors. Also, clear policies should have been established and agreed to by various stakeholders, and critical process control instructions or standard operating procedures (SOPs) have been reviewed and their design must eliminate or minimize process risk. In addition, supervisors and employees must know and use the standards. To be effective, employees and supervisors must be provided with the right tools, procedures, and equipment to do the job correctly the first time.

Planning moves management from good intentions or wishes to the greater challenge of reality and improving the existing position of the organization, including quality and environmental health and safety (QEH&S).

IMPLEMENTATION AND OPERATIONS

Management must consistently consider QEH&S policies in planning, implementation, evaluation, documentation, and corrective/preventive actions. This program could ultimately include planning for the use of contractors, subcontractors, and suppliers. In addition, this commitment should be included in the company's mission statement and management philosophy and further reflected in its policies and corporate/facility goals and objectives. Management's quality and EH&S plans should be discussed in the annual report and established as an integral part of management–supervisor training and evaluations. It should also be included in management meeting agendas, including progress reviews or evaluations of previously established objectives.

For effective implementation, organizations must develop the capabilities and support mechanisms necessary to achieve its QEH&S policy, objectives, and targets. The policy must also be clearly communicated to all managers, supervisors, and employees, prior to implementation, and will require an evaluation tool or tools to monitor performance toward achieving goals and objectives. Performance monitoring should clearly identify the necessity to change plans or establish new objectives along the road of continual improvement.

> "The expectations of life depend upon diligence; the mechanic that would perfect his work must first sharpen his tools."
>
> —*Confucius*

There are specific steps to take and a logical process for implementing your integrated management system (IMS). QEH&S Management Systems must include processes for setting policy and process standards. These standards are for managing operational risk reduction and quality enhancement, including regulatory compliance. However, processes must have a performance versus regulatory focus. Well-managed systems generate compliance, whereas management focus on regulatory adherence tends to generate incomplete and costly compliance, oftentimes contributing little to operational efficiency. However, regulatory

Figure 10 *The Path.*

compliance should not be ignored. It simply should not be the overriding focus of the management system.

Facilities must have the resources and procedures in place to identify and access applicable legislative or regulatory requirements, including a periodic review process to ensure that changes are identified, understood, and considered in the planning process as appropriate. Key considerations regarding the establishment of a policy and procedures manual could include the following steps.

- Develop procedures and process design review, including the adequacy of tools, equipment, and SOP's and their maintenance to ensure safe quality work performance.
- Identify needs of and implement training and communication process for supervisors and employees.
- Establish performance expectations and appraisal processes, and encourage employee participation in the process (to be effective this process should be as objective as possible) subject to review, and ultimately resulting in the sharing of the organization's profits (see comments about Lincoln Electric in Chapter 7 and Appendix III, they are a very successful example of this process).
- Develop accident or nonconformance investigation procedures (see Chapter 8, which includes requirements to identify the multiple root causes of the

nonconformance and a plan to correct/prevent reoccurrence), this process should include follow-up of previously identified corrective/preventive actions resulting from audits or accident and nonconformance investigations.

- Audits, inspections, and routine monitoring or job observations are the fuel for continuous improvement, allowing for reinforcement of desired activities, modifications, as appropriate, of existing policies or procedures and the immediate correction of non-conformance.
- Medical surveillance and occupational health, including employee wellness, return-to-work programs, and strong claims management support.
- Establish system for documentation of all aspects of the process enhancement effort, including periodic senior management review.
- Risk identification and assessment/control: Process risk ranking, prioritization, and follow-up must specifically address design and procurement control, including tools, equipment, and facility use and maintenance and contractors/vendor procedure control, including service and product delivery, material handling, storage and transportation, waste minimization, training, and supervision.

Other requirements would include consideration of emergency response and contingency planning in the facility reviews and design of the management system.

Risk assessments should be rigorous and should follow a systems safety process, such as Management Oversight Risk Tree analysis (MORT), Failure Mode Effects Analysis (FMEA), and Change Analysis; flow charting should also be used to help identify, evaluate, prioritize, and control quality and EH&S risks of facilities and operations, including legislative or regulatory requirements.

Community and other stakeholder's views should be part of the risk assessment process. Review prior incidents and the results of their evaluations to help determine the efficiency and effectiveness of existing processes, and the resources committed to their management enhancement process.

- Clear assignment of responsibility and accountability and, to the extent possible, accountability should be integrated in the QEH&S management systems policies, procedures, and instructions.

Process Design and Systems Safety Review

Table 9 at the end of this chapter is a simplified systems safety check sheet intended to help you review readiness to implement new operational changes to procedures. Its intent is to categorically walk you through the significant resource inputs to a process to ensure the facility or organization's readiness to execute change or to introduce new processes without generating uncontrolled risk for employees, the environment, or product/service delivery. The review consideration is focused on things like the adequacy of tools, equipment, and facilities and their maintenance, communications, including training, procedures and supervision/management of the process.

The concept in this table goes back to management oversight risk tree analysis (MORT) and other systems safety and Hazard-Operability (HAZ-OP) tools that were developed primarily for the Department of Energy, the U.S. Navy Nuclear Power Program, and NASA with design input from companies like EG&G. A systems safety review can be very complex and labor-intensive to complete, or it can be a relatively easy process of defining processes for job-task observations, or group-facilitated qualitative risk assessments as discussed in the Chapter 5. Dr. Deming's "fish bone" diagrams can also assist in systems safety reviews of critical processes.

Do not confuse systems safety with employee safety, although systems safety should consistently be part of the review process. Systems safety and risk assessments look at a broad range of processes or operational risks to identify and determine the most effective way to eliminate or control the risk of nonconformance. Nonconformance in turn can be interpreted to include product/service quality, rework, safety, or environmental risk.

Procurement—Including Contractors Certification Process

The goal of effective procurement policies is to ensure that design and customer requirements are met through clear purchasing processes and documentation and evaluation of vendor capability to meet product or service delivery specifications. Price consideration should be considered only after design or product specifications are ensured. Procurement procedures are also intended to ensure that risks are minimized with purchasing, shipping, storage and the use of hazardous materials ("cradle-to-grave" material management). Procurement procedures must ensure that products and services meet the specifications defined by the customer or that they are part of the product/service design criteria or operations requirements.

Procurement requirements include considering the substitution of hazardous for less hazardous material but to still allow operations managers and customers to achieve their production or design requirements. Substitution of materials, however, should be made only if it is approved by the "stakeholder/designer" requesting the material and the EH&S manager evaluating the material's potential risks, including shipping, storage, and containment requirements. In addition, purchasers of hazardous materials must consider any changes or impact to regulatory requirements with regards to hazardous material procurement. These factors should include changes to their emergency response plans and the notification of community support services such as the fire department. Procurement policy and procedures should include a process for qualifying vendors and the development of contractor- and vendor-approved lists. The approval process, vendor certification, and product-service delivery inspection processes should be documented to ensure that vendor supplied materials and services meet QEH&S requirements.

Initial QEH&S Review (Gap Analysis)

Every organization will find that some elements of a QEH&S management system are in place. What is less common is the linking of these elements into a coordinated overall system for improvement, such as integration with other management process control systems for quality and/or environmental management.

A useful starting point is to critically compare the basic intent of each element in the management system guidelines, policy, and regulatory requirements, and the management practices and procedures that are currently being used. Many organizations have obsolete procedures and need to compare the system and policy requirements with what actually occurs to obtain a realistic assessment of the implementation plan and integration task requirements.

Core elements that could be focused on initially:

- Clear management responsibility for QEH&S
- Hazard identification, risk assessment, process change/improvement, and corrective action
- Documentation of critical procedures
- QEH&S inspections, performance observations, and job-task analysis of critical procedures, facilities, tools, and equipment
- Communications, including training, procedures, and supervision (Are communications clear, consistent, and appropriate?)

The initial review will establish a base line or starting point for the development and implementation of a management system. The management system audit compares the current situation in the company with each element of the proposed requirements or policy (Gap Analysis). What gets measured gets done, and usually with improved quality; therefore, tools such as audits and job observation are critical.

It should be recognized that there might be significant short falls or elements, which may not exist currently in the organization. The audit, however, will identify these elements and make corresponding recommendations to deal with a system of non- or partial conformance. In addition, a culture review may be necessary to evaluate actual and perceived program commitment and execution.

Common techniques and tools for initial review:

- Questionnaires
- Interviews with employees
- Checklists
- Direct inspection, observation, and measurement
- Audits
- Record review
- Benchmark with similar organizations

Other useful sources of information can be found in records such as nonconformance reports or product rejection/rework reports, illness, incident, and first-aid reports kept by the facility. Insurance companies are also able to provide feedback on an organization's or industry's claims experience and the breakdown of the components of the insurance premium and how it compares within an industry group. In addition, information on the types of losses experienced in many insured areas of the business could provide insights into transportation and security risks as well as property and life safety risks at the facility and industry levels. This type of information will be very useful in conducting both risk assessments and in establishing process improvement benchmarks.

HUMAN, PHYSICAL, AND FINANCIAL RESOURCES

The size of organizations may impose constraints on implementation. To overcome these constraints, external QEH&S resources can be utilized. Such resources might include the following.

1. Shared technology and experience from larger client organizations—cooperative approaches to develop industry, employer, or trade association best practices
2. Assistance from government QEH&S organizations
3. The use of consultants and the collective engagement of consultants
4. Advice and training from suppliers
5. Assistance provided by insurance carriers
6. Attendance at QEH&S seminars and/or professional associations
7. Support from university research centers or government agencies.

INTEGRATION

An organization that has a documented quality and/or environmental management system should find that integrating QEH&S saves time and resources and improves the overall management system. This objective may be accomplished in some cases by integrating process enhancement objectives in specific departments or processes versus facility-wide execution. Organizations should work towards obtaining the capacity to balance and resolve conflicts between quality and EH&S, as well as other organizational objectives or priorities; and aligning and integrating EH&S into the overall business management process, for example, systems including information systems conforming to the ISO 9000 or 14000 series of standards for integrated reporting and documentation.

EMPLOYEE INVOLVEMENT AND CONSULTATION

Process improvement is most effective when people from all levels of the organization are encouraged to participate in the development and implementation of the system. People are more likely to embrace change if it is not imposed. Involving employees in change decisions and responding to people's concerns will help establish common goals between managers and employees. Employees and different management "stakeholders" will need to be involved in performing process risk assessments. This need is especially true for identifying organizational and process risk and ranking the risk and proposed controls to either eliminate or minimize the risk. Examples of employee involvement include the following.

- Participating in joint labor-management committees, PET
- Conducting site inspections, job-task analysis, and audits
- Performance management or behavioral-based safety job observations and task analysis
- Analyzing risks and proposed control strategies
- Preparing work procedures to eliminate or reduce process risks and work hazards
- Developing and revising QEH&S rules and procedures
- Training new and current employees
- Providing programs and presentations at QEH&S meetings
- Conducting accident incident investigations
- Participation in decision making throughout the organization's operations

COMMUNICATIONS

Effective communication is an essential element of the QEH&S management system. Communications includes training, procedures, supervision, and management. Lack of effective communications is the quickest way to establish substandard process conditions and process breakdown. Too often conveyances are lost between the various aspects of the communications process by the time they reach the employees developing the products and services for the customer. Unless communications are clear and aligned they will significantly contribute to an environment that fosters substandard performance.

Providing appropriate information to the organization's employees, their representative, and other stakeholders serves to motivate and encourage understanding and acceptance of the organization's efforts to improve performance. In addition the system must ensure that applicable QEH&S instructions/communications are available and that out-of-date documents are removed from service.

Employees must be encouraged to communicate, identify, and report QEH&S process concerns. In addition process failures or incidents must be investigated to determine the multiple root causes in the system that contributed to the process failure. The lessons learned from these evaluations must then be clearly communicated back to applicable staff and management as a key step in the corrective/preventive action process.

Internal Reporting Communication Procedures

Internal reporting procedures need to be established for reporting the following instances.

- Product nonconformance
- Non-conforming tools, equipment, materials, or procedures
- Accident/incident occurrences
- Process nonconformance
- Health and safety performance
- Hazard identification
- Environmental releases or nonconformances
- Process changes

Organizations should have procedures for ensuring that pertinent QEH&S information is communicated to all staff, management, and contractors. In large organizations, this task will require good planning, especially for multinational companies confronted with language and cultural issues. Effective communications are critical for sharing "lessons learned" as a result of an accident or process nonconformance study and for successful communication of changes to process design, materials, or procedures. The communication system should also consider other external "stakeholders," for example, customers, industry groups, and the community and regulators and their need to know of the changes or "lessons learned" as a result of a process failure investigation (see Chapter 8).

Communications require arrangements to determine information needs are met by the following.

A. Communicating the results from management systems, monitoring, and audit review to those within the organization who are responsible for and have a stake in the organization's performance

B. Identifying and receiving relevant QEH&S information from outside the organization
C. Ensuring that critical information is effectively communicated to people inside and outside the organization who require it
D. Establishing a QEH&S manual that formerly establishes policy for incorporating work procedures and general QEH&S requirements (The manual should include all relevant programs, especially those that parallel the QEH&S. It should also be available at locations where operations essential to the effective functioning of the QEH&S system are performed.)

Commonly used methods of internal communication:

- Newsletters
- Bulletins
- Notice boards
- Tailgate or lunch box meetings with supervisors, or other team briefings
- Management meetings
- Videos
- Hard copy or electronic mail
- Signs

Commonly used methods of external communication:

- Annual reports
- Publications
- Inserts in industry publications
- Paid advertising
- Telephone inquiry services
- Submissions to government

PROCEDURES: PROCESS CONTROL INSTRUCTIONS (PCI) OR STANDARD OPERATING PROCEDURES (SOP)

The logical first step toward working more safely and with improved quality is to develop written job procedures or practices. These practices become teaching tools and standards to measure performance when conducting job observations, incident investigations, or evaluations of product process failures. Keeping it simple and relevant is critical to having an effective instruction.

To accomplish a goal of having simple, clear, and easy-to-understand process instructions you must first understand the process and its inherent risks. Chapter 5 discusses risk assessment processes, including process flow charting and identifying those steps in the process that have a history of nonconformance or a potential for process failure. When you have a clear understanding of the process as it is, and how it should be, design changes or the development of meaningful (SOP's) become possible, and necessary to ensure conformance.

Reporting

Effective reporting should cover the positive initiatives the organization is taking to identify hazards and control risks and can include the following steps.

a. Critical task process conformance
b. Reports of levels of conformance with procedures
c. Reports on performance against targets
d. Reports on improvements made
e. Reports on underlying reasons for incident occurrences (multiple root causes)
f. Reports on results of audits, and lessons learned from process changes, or nonconformance investigations or critiques

> The benefits of better quality through improvement of the process are thus not just better quality, and the long-range improvement of market position that goes along with it, but greater productivity and much better profit as well. Improved morale of the work force is another gain: They now see that the management is making some effort themselves, and not blaming all faults on to the production workers.
>
> —*W. Edwards Deming,* Quality, Productivity, and Competitive Position

Maintenance

In 1992, the U.S. Department of Commerce organized a trip to East Germany for my partner, myself, and the heads of 10 other U.S. environmental companies. I spent 10 days there with my partner meeting German engineering firms and visiting environmentally contaminated sites. Both the American and German governments were interested in trying to create joint venture (JV) opportunities with East German engineering firms. Their objective was to cost-effectively initiate the environmental remediation of the former Soviet industrial sites as a step towards encouraging western investment in East Germany.

A visit to Eastern Europe or the former Soviet Union at that time, would likely have changed anyone's opinion that communism or centrally controlled systems work. I have never seen so many facilities with such obvious and serious environmental problems as the facilities we visited, notably the properties in "Bitterfield" and some facilities in Rostok located in northeastern Germany on the Baltic Sea.

I am convinced that not only did these facilities have serious environmental problems, but I am also sure that their operations generated significant safety and health risks for the employees in the facility, their families, and the surrounding communities. It is difficult to believe that these operations could efficiently produce quality products that would satisfy the needs of customers.

One of the key root causes for these obvious process failures was an ineffective facility tool and equipment maintenance system. In fact, we visited one city where only 60% of the community was connected to the wastewater treatment facility, and more than half of the facilities treatment capacity was shut down because of the failure of a single pump. Without defined procedures and maintenance critical processes were likely to fail, while also putting the employees at risk. Any substandard performance of the employees at these facilities would be the direct result of substandard conditions generated by the lack of a defined efficient operating system, tools, equipment, etc.

All critical operations, facilities, tools, and equipment must be inventoried and included in a routine but defined maintenance schedule. Without this process, the issue is not a matter of if there will be rework, loss of product, injuries, or environmental problems but when and how severe. In reality, as illustrated in Chapter 1, operations will probably have hundreds of minor nonconformances before these incidents become glaringly obvious in the form of a serious incident, loss of product, or even the business. Management must recognize these incidents or "red flags" and determine their multiple root causes before they have a more serious manifestation of the nonconformance.

MANAGEMENT AND CONTROL OF CONTRACTORS AND VENDORS

Organization should evaluate and select contractors and vendors on the basis of their ability to meet Quality and EH&S requirements. Procedures for qualifying vendors and contractors must be established and documented and should not be based solely on price. The contractor or vendor's ability to document existing QEH&S systems within their own company is the most critical qualification standard.

This standard is closely followed by a documented history of successful QEH&S performance, including product/service conformance to requirements. Organizations should establish and maintain documented procedures for controlling QEH&S aspects of contractor work and vendor delivery of products or materials. These procedures should cover the following steps.

- Pre-planning for medium and long-term contracts, this step will involve carrying out a full QEH&S pre-qualification/evaluation processes; QEH&S aspects for short-term contracts should be checked by questionnaire or review.
- Defining responsibility and accountability for communication links between appropriate levels of the organization and the contractor prior to contract commencement.
- Site safety rules, including applicable EH&S regulatory requirements.

- QEH&S training of the contractor personnel where necessary before commencement of work.
- Monitoring QEH&S aspects of contractor activities on site.
- Communication of accidents and incidents involving the contractor's personnel.
- Removal of hazardous waste generated by the contractor's activity.

EMERGENCY PREPAREDNESS AND CONTINGENCY PLANNING

A distinction needs to made between emergency response and contingency planning. Contingency planning is generally much broader in operational scope. Emergency preparation is in most cases a subset of contingency planning efforts. Contingency planning covers the identification and management of all operational risk, including financial risk. Emergency planning and response procedures are what the name implies and include plans that most likely require involvement of community response personnel, fire departments, hospitals and emergency medical personnel, the police department, and possibly outside contractors.

Contingency planning may interface with or have aspects of emergency response and could be included in an emergency response planning process. However, contingency planning is really looking at the "what if" of organizational or business risk. If we identify a potential change that would cause a risk, what actions will be taken to either prevent or minimize its impact? Contingency planning managerially or procedurally is like redundant engineering controls, or back-up systems in the event of a primary system or process failure, or as a way to identify risks and execute process changes to eliminate the need for a contingency plan.

Both emergency and contingency planning involve preparing written plans for every facility, outlining the steps to be taken in the event of a variety of emergency situations, or plans to be executed in the event of a business or systems change that requires the operations manager to activate the contingency plan or back-up system.

Emergency plans are not a fragmented series of plans but an overall process with provisions for each type of emergency. Emergency preparedness, hazardous materials business, and communications plans are related and should be integrated into the plan as appropriate.

QEH&S MANAGEMENT SYSTEM DOCUMENT CONTROL AND RECORD-KEEPING PROCESSES

Documentation and record-keeping are essential for preventing unnecessary losses. Records can also become the basis for sharing "lessons learned" from nonconformance or accident investigations and can be a tool for the development of more effective procedures and processes.

Facilities should to the extent possible simplify the records and documentation process. One method is to take advantage of intranet systems. ISO auditors too often cause companies to generate excess paper, documenting rather than focusing on the management systems and the continual improvement process. The primary focus of managers should be on improving work processes or systems integral to the quality and environmental health and safety aspects of their business. Documentation and record-keeping is a key part of any management system but should be secondary to process management and tailored to the needs of the organization, not the driving focus of the system. Documentation should include, to the extent possible, applicable regulations and documentation compliance.

Operational processes and procedures should be defined and appropriately documented and updated as necessary. The organization should clearly define the various types of documents, which establish and specify effective operational procedures and controls. Records should also include reports and assessments of accidents or other process failures or nonconformances. These reports should document "lessons learned" by identifying the multiple root causes of the process failure. Documentation of corrective actions and sharing of lessons learned are key to process improvement and must be documented.

QEH&S management systems documentation supports employee awareness and enables the management review of the systems. The degree and quality of the documentation will vary depending on the size and complexity of the organization and the level of QEH&S management systems' integration. Organizations' should consider maintaining a documentation summary or table collated with the QEH&S policy, objectives, and targets and the "shall" documentation requirements of ISO 9001 or 14001. Documentation should include the following,

- The means of achieving QEH&S objectives and targets
- Documentation of key roles, responsibilities, and procedures
- Direction or related documentation describing other elements of the organization's management system, where appropriate
- Demonstration that the applicable management system elements are implemented
- Minutes from QEH&S management review meetings

A complex range of information can result from management system documentation efforts if they are not carefully planned and organized. The effective management of these records is essential to the successful implementation of the management systems. The key features of good QEH&S information management include means of identification, collection, indexing, filing, storage, maintenance, retrieval, retention, and disposition of QEH&S management systems documentation and records.

The QEH&S management system should be described in a manual that references related documentation and the requirements for and maintenance of management system records or records dictated by or associated with all related processes or procedures, or as required by regulation. Procedures should be periodically reviewed and modified as necessary to conform to process changes, facility, and regulatory, customer, or industry standards or policy.

Processes should be in place for the review-approval and "sign-off" of all revised control procedures. Current versions of all relevant documents must be available and in locations convenient for personnel or "stakeholder" access as necessary to effectively conform to management system requirements.

Document retention times should be determined and an indexed system established for record maintenance and retrieval, including the removal and replacement of obsolete documents (e.g., management, system policy, and manuals, work procedures such as process control instructions or SOP's, EH&S regulations, audits, inspections and calibration records, job observation records, maintenance, medical, personnel and training records). Individual documents should include draft or version number and execution dates.

Facilities should also maintain archives for specific records required by regulation such as EH&S-related procedures. Documents should be identified with revision numbers and date in the footer and a list or inventory of all control documents should be included. A process for removing obsolete documents and procedures (especially work process control instructions) and replacing them with current documents will also be necessary. Facilities should maintain these archives for specific required records.

Four Tiers of Documentation in ISO 9000

- Quality and policy.
- Quality system procedures.
- Product and process specifications, inspection instructions, process parameters, testing procedures, packaging and labeling requirements, delivery schedules, preventive maintenance, analytical testing, process and equipment calibration procedures, and work instructions for all activities affecting customer products and services.
- Records of all activities affecting quality and proof that the internal audit/corrective action program is documenting continuous improvement.

Table 8, below, is a brief overview of the ISO record-keeping requirements.

The QEH&S management system should be described in a manual that references related documentation and the requirements for and maintenance of management system records, or records dictated by or associated with all related processes or procedures. Procedures should be periodically reviewed and modified as necessary to conform to process changes, facility, regulatory, customer, or industry standards, or policy.

All records necessary to document conformance with the facility's management system, regulatory or permit requirements, QS, 9000, the Chemical

TABLE 8. ISO-9000 Requirements for Procedures and Records

Requirements for Documented Procedures		Requirements for Records	
4.3.1	General contract review	4.1.3	Management review (quality system)
4.4.1	General design control procedures	4.2.3(h)	Identification and preparation of quality records (see 4.16)
4.5.1	General document and data control	4.3.4	Records (contract review)
4.6.1	General purchasing	4.4.6	Design review
4.7.1	Control of customer-supplied product	4.4.7	Design verification
4.8	Product identification and traceability	4.6.2(c)	Establish and maintain quality records of acceptable subcontractor (see 4.16)
4.9(a)	Process control: documented procedures defining the manner of production, installation, and servicing when the absence of such procedures could adversely affect quality	4.7	Control of customer-supplied products
4.10.1	General inspection and testing	4.8	Product identification and traceability
4.11.1	General control of inspection, measurement, and test equipment	4.11.1	General control of inspection, measurement, and test equipment
4.13.1	General control of nonconforming product	4.11.2(e)	Control procedures-maintain calibration records for inspection, measurement, and test equipment
4.14.1	General corrective and preventive action	4.9	Process control
4.15.1	General handling, storage, packaging, preservation, and delivery	4.10.2.3	Receiving inspection and testing-incoming product released for urgent production purposes prior to verification is positively identified and recorded
4.16	Control of quality records	4.13.2	Review and disposition of non-conforming product
4.17	Internal quality audits	4.17	Internal quality audits
4.18	Training	4.18	Training
4.19	Servicing	4.14.2(b)	Corrective action
4.20.2	Statistical techniques, procedures		

Manufacturer's "Responsible Care," or ISO 9000/14000 requirements must be maintained and archived for easy identification and recall as necessary.

However, as mentioned earlier, don't let the auditor's documentation needs drive the process. Let the needs of the process or system define the level of effort applied to documentation. In addition we should rely on our IS systems to the extent possible to simplify documentation and record-keeping. Focus on process improvement, and let the necessary documentation flow from it rather than trying to build the system around audits and required documentation.

Records

Organizations should establish and maintain procedures for the identification and maintenance of management system records essential for the successful implementation and documentation of the QEH&S management systems. Effective information management includes these key features.

- Identification
- Collection
- Indexing
- Filing
- Storage
- Maintenance
- Retrieval
- Retention
- Disposition of pertinent management systems documentation and records
- Confidentiality of personnel medical records.

Appropriate document retention times may vary due to regulatory requirements. Control documents must be identified and a system established to ensure that all control documents are current. A master list of control documents can ensure identification of those current revisions and document changes. including change dates. Record-keeping provides benefits for legal defense program reviews and, if done smartly using today's IS technologies, the process can reduce cost for sharing information and documenting change and the improvement process between facilities and work centers.

Records should include the following elements.

- External (e.g., legal) and internal (i.e., QEH&S performance) requirements
- Work permits
- QEH&S risks and hazards
- QEH&S training activity
- Inspection, calibration, and maintenance activity
- Data monitoring
- Incident details, complaints, and corrective/preventive follow-up actions
- Product or service identification and specifications
- Supplier and contractor information
- Audits, reports, and documentation of corrective/preventive actions

Action Plan

- Evaluate Figure 3 at the end of Chapter 1, Management System Implementation Process, tailor local facility/business processes, and perform initial GAP analysis.
- Develop process control instructions/SOPs for critical operations/tasks, including QEH&S requirements.

- Establish procurement/vendor review and approval process.
- Establish systems safety design review of new and established critical processes.
- Establish training program.
- Develop an inventory of critical operations and establish schedules for qualitative risk assessments.
- Conduct initial audit (Gap Analysis), and schedule annual follow-up audits (Note: This step could also include a cultural climate review survey).
- Establish maintenance schedule for critical tools and equipment.
- Develop waste minimization and recycling plans.
- Establish emergency response plan.
- Develop maintenance schedule for critical parts and equipment.
- Establish control document, record-keeping procedures.
- Develop operations contingency plans.
- Identify business risk and prepare operations contingency plans.

Involve human resources so that the stress of the organizational and process changes are mitigated and that change is clearly and positively communicated at all organizational levels

PROCESS RISK ANALYSIS OF PROCESS INPUTS

Tables 9a–9h are simple systems safety-type process review check-sheets. They can be used to help you evaluate new process readiness/completeness or risk. They can also be used to access existing systems or processes, or process nonconformance as defined in the revision of these check-sheets in Chapter 8 (See Table 16).

TABLE 9a. Personnel Selection, Training, and Mentoring: On-the-Job Training, Supervision. Can a Person's Lack of Physical Ability, Skill, or Knowledge Contribute to Process Nonconformance?

Yes	No	Potential Nonconformance Causes:
		1. Are personnel given a pre-placement medical examination to determine physical qualifications for the job?
		2. Can the task or process be simplified?
		3. Do pre-placement interviews and orientation include a review of the work conditions and requirements such as the handling of hazardous materials or working in high places or confined spaces?
		4. Our employees given a general QEH&S work-site orientation?
		5. Is on-the-job training provided? Is it renewed in the last year with any kind of updated process-control training, briefings, or written material?
		6. If the employees have a lack of aptitude, are workplace modifications made to compensate for their capabilities?

TABLE 9a. (*continued*)

Yes	No	Potential Nonconformance Causes:
		7. Are employees given adequate supervised practice to develop necessary job or task skills?
		8. Are tasks performed frequently enough to maintain skill levels and familiarization with procedures? Are there other training needs?
		9. Do the hiring, placement, and training programs include adequate examination and evaluation of each person's basic capability to perform the job safely and of IAW customer quality specifications? Do these programs include provisions to modify the workplace or the activity to accommodate personnel with permanent or temporary disabilities?
		10. Are there adequate supervised demonstrations and practice sessions to develop skills and procedural knowledge?
		11. Is there additional loss potential due to inadequate standards or compliance?
		12. Do the employee training, personal communications, and group meetings include adequate training to develop and update each person's knowledge of how to perform in conformance with QEH&S requirements?
		13. Are personnel training standards adequate and in compliance?
		14. What additional loss potential exists due to inadequate personnel selection, training, and supervisor standards?

TABLE 9b. Personal or Job Stress: Can a Person's Physical or Mental Stress be a Factor or Contribute to Process Nonconformance?

Yes	No	Potential Nonconformance Causes:
		1. Do we have the ability to identify a person suffering from temporary illness or stress, such as family or personal problems or decreased abilities, including alcohol, drugs, or controlled medications?
		2. Is the task load, pace, or duration high enough to physically or mentally fatigue the employees? Are process controls threatened by production demands?
		3. Are employees exposed to environmental stress from: – vibration – noise – heat or cold – chemicals or contaminants
		4. Will physical movements be constrained or are abnormal body positions required?
		5. Will difficult judgments be required while doing the task?
		6. Are there activities or conditions that might cause distractions?
		7. Are environmental controls, engineering controls, and job observation practices adequate to identify and control the effects of stress-producing situations?
		8. Are standards adequate to define the control measures and the level of performance needed to identify exposures and control the risks, even with increased production demands?
		9. If existing standards are adequate, are they in compliance? If not, why?

TABLE 9c. Engineering: Is the Design of the Production Facility, Process, or Equipment Adequate?

Yes	No	Potential Nonconformance Causes:
		1. Are QEH&S hazard risk assessments made during the concept design stage of the process?
		2. Are operation effects that interface with the process considered in the QEH&S risk assessment or hazard analysis?
		3. Are all appropriate standards, codes, and previous nonconformance reports and lessons learned researched to establish criteria for the process or facility design?
		4. Are controls for the identified risk incorporated into the final design?
		5. Will the QEH&S staff participate in the design review process? If not, why not?
		6. Will a systems safety review (HAZ OP, etc.), be performed with design and production engineering?
		7. Have the QEH&S criteria been specified to the contractor?
		8. Will pre-construction or process installation conferences be held with contractors and facility employees/supervisors, and will they include an agenda requiring QEH&S specifications and readiness review?
		9. Will the facility or equipment be examined to ensure that QEH&S criteria are followed before acceptance?
		10. Will modifications to the original design be evaluated for QEH&S risk during the design change process?
		11. Are QEH&S standards or regulations considered? Is past experience evaluated, and are appropriate criteria incorporated into the design and construction of the product–production life cycle process
		12. Is there additional loss potential due to inadequate engineering, standards, or compliance?

TABLE 9d. Materials and Procurement: Can Inadequate Purchasing, Management of Materials, or Disposal of Scrap, Waste, or Obsolete Items be a QEH&S Process Conformance Issue?

Yes	No	Potential Nonconformance Causes:
		1. Are EH&S standards and quality specifications researched and specified to vendors of tools and equipment or to other service providers?
		2. Is the coordination of requests for tools, equipment, or material purchase adequate, and does it include a review by QEH&S professionals and managers?
		3. Will QEH&S information and instructions on equipment or materials be communicated to the area or work center supervisor for approval/concurrence?
		4. Are the hazardous material labels on materials and safety placards on tools kept intact and legible? Do labels or other markers include periodic calibration, testing, maintenance due dates or other requirements for the equipment/tools?

TABLE 9d. *(continued)*

Yes	No	Potential Nonconformance Causes:
		5. Have appropriate handling methods and personal protective equipment been determined for the loading, moving, storing, and issuing of hazardous materials?
		6. To prevent deterioration or instability, are materials and equipment stored properly prior to use, including secondary containment, material segregation, and data inventory for first in-first out (FIFO) material management?
		7. Are the hazardous properties of waste, scrap, and unserviceable or obsolete equipment identified so disposal methods could be initiated?
		8. If these elements are included in the QEH&S program, are they adequate to ensure that potential risks are identified before items are purchased or facilities are constructed? Are risks controlled throughout the facilities, equipment, and material life, including periodic testing and calibration?

TABLE 9e. Maintenance (Facility, Equipment, and Tools): Is Premature Failure or Malfunction of Equipment or Facilities a Process Risk?

Yes	No	Potential Nonconformance Causes:
		1. Have service manuals and government codes been checked to determine the needs for preventive maintenance requirements?
		2. Has a maintenance schedule been prepared?
		3. Are adequate time and priority given to preventive maintenance?
		4. Is all preventive maintenance conducted according to schedule?
		5. Are written maintenance instructions given to work center supervisors and employees?
		6. Is preventive maintenance work regularly audited?
		7. Are lubrication and service points readily accessible?
		8. Are the specified lubricants or approved substitutes used?
		9. Are parts stored, delivered, and assembled correctly?
		10. Are adjustable parts set to correct control specifications?
		11. Are assembly and adjustment of parts made by using clearly written standards in technical manuals or SOPs?
		12. Are parts cleaned and resurfaced as required?
		13. Are defects or deficiencies documented in written repair work orders?
		14. Are maintenance personnel given adequate repair or maintenance instructions?
		15. Are adequate time and assistance scheduled for the repair or maintenance work?
		16. Is equipment checked when the repair or maintenance is completed?
		17. Does a skilled craftsman or mechanical inspector check the repair?
		18. Are specified parts or approved substitutes used?
		19. Do the engineering purchasing controls and inspections include research of requirements for maintenance?

TABLE 9e. (*continued*)

Yes	No	Potential Nonconformance Causes:
		20. Do the job analysis, procedures, and personal communication include development and communication of maintenance instructions?
		21. Are standards to define control measures and the level of performance needed to identify loss exposures and control risks
		22. Are these standards in compliance?
		23. What is the additional loss potential due to this lack of compliance?

TABLE 9f. Process Control Instructions (PCIs)/Standard Operating Procedures (SOPs): Are Adequate Methods, Procedures, Practices, SOPs, or Rules Used?

Yes	No	Potential Nonconformance Causes:
		1. Are there QEH&S methods for the process developed, and are these methods communicated to all responsible managers, supervisors, and employee trainers?
		2. Are procedures correct and current? To the extent applicable, are the QEH&S requirements incorporated into existing SOPs?
		3. Have the procedures been reviewed and updated?
		4. Do the procedures take into consideration risk assessments?
		5. Are lessons learned for the general type of work involved in previous incidents available to supervisors and employees?
		6. Do the engineering controls, job analysis procedures, and organizational rules include study of work processes and methods development?
		7. Are the standards adequate to ensure that methods, procedures, practices, or rules are used as appropriate for the hazard and the evaluated degree of risk?
		8. If the program and standards are adequate, are they in compliance?
		9. What is the additional loss potential due to inadequate procedures or compliance?

TABLE 9g. Management and Supervision: Are Leadership, Management, and Supervision Adequate?

Yes	No	Potential Nonconformance Causes:
		1. Is there a written QEH&S policy, and is it understood by the managers and employees in the area involved?
		2. Is control of the loss exposure affected by increased production demands and limited budgeting of personnel, equipment, materials, or other financial or resource support, including communications?
		3. Is responsibility for QEH&S assigned to a specific work center person or position, and is sufficient authority given to enable effective control?
		4. Are responsible persons trained in QEH&S related to the type of process and risk exposures?

TABLE 9g. (*continued*)

Yes	No	Potential Nonconformance Causes:
		5. Are potential QEH&S risk exposure and its control measures discussed at management meetings?
		6. Does the QEH&S program reference manual provide adequate guidance to prevent the risk of nonconformance?
		7. Are QEH&S program control measures that can prevent the loss exposure risk audited?
		8. Are the management training elements of the program adequate to ensure that planning, organizing, leading, and controlling of QEH&S are effectively incorporated into the organization management program?
		9. If the program and standards are adequate, are they in compliance?
		10. What other loss potential exists due to inadequate program, standards, or compliance?

TABLE 9h. Tools and Equipment: Are Suitable Tools, Machines, or Vehicles Available?

Yes	No	Potential Nonconformance Causes:
		1. Are the tools used in activities related to the process correct for the task, as specified in instruction manuals and safety standards?
		2. Are tools used correctly, as anticipated when purchased or built?
		3. Are the tools or equipment in good condition and not worn beyond limits for safe use?
		4. Are the equipment, tools, or facilities used within an expected service life?
		5. Are adequate numbers of correct tools and parts in condition and available for the task?
		6. Can the equipment be operated at excessive speed, temperature, or any other abnormal way?
		7. Are tool issue or storage areas convenient and well-organized so the proper tools could be readily obtained?
		8. Contractor or others who are not adequately trained are not allowed to use the facilities, tools equipment.
		9. Are inspection frequencies increased as the tools, equipment, or facilities became older or their use rate increased?
		10. Is there a procedure for prompt repair, adjustment, sharpening, and other maintenance of tools and equipment?
		11. Have defective tools been replaced and salvaged, made inaccessible, to prevent unintentional use until refurbished, or replaced?
		12. Is defective equipment shut down, locked out, and tagged to prevent its use until it is repaired or replaced?
		13. Are equipment defects reported to supervision?
		14. If these activities are included in the program, are the standards adequate to insure that safe tools and equipment are used?
		15. If the program and standards are adequate, are they complied with?
		16. What other loss potential exists due to inadequate program, standards, or compliance?

5

Risk Assessment Process

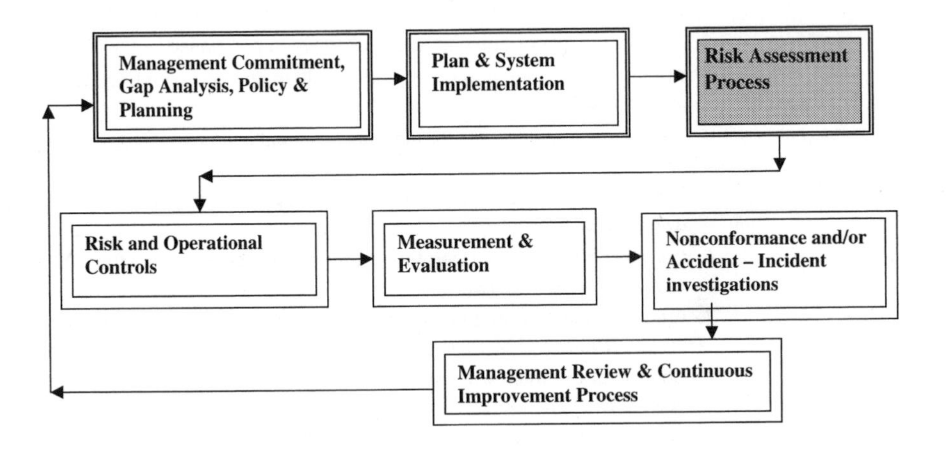

Risk and quality management are simply two sides of the same coin.
—*Mark Andrews, Liberty Risk Services, Melbourne, Australia*

As I have mentioned before, I do not believe that it is possible to make all process improvement changes at once. Organizations have limited resources and can only absorb so much change at one time. Therefore, you will need a process to identify facility risks and a method to prioritize these risks and their proposed controls. A number of tools are available for performing risk assessments. In most instances, however, a group-facilitated qualitative risk assessment is all that will be needed for risk and proposed control/resource ranking.

In this chapter we briefly outline the qualitative risk assessment processes and identify some key group facilitation procedures. However, a better text to help you with your group facilitation skills is *In Search of Solutions: Sixty Ways to Guide Your Problem Solving Group*, by David Quintan-Hall and Peter Renner (1940).

Quality circles, and process enhancement teams benefit from and oftentimes use nominal group techniques to identify and prioritize process risks and to develop various control alternatives, including process change.

EVALUATING BUSINESS AND ORGANIZATIONAL QUALITY AND ENVIRONMENTAL HEALTH AND SAFETY (QEH&S) RISK

Organizations should adapt the following practices.

- Establish and maintain documented procedures for the identification and assessment of QEH&S risks.
- Establish and maintain documented safe work practices (or equivalent procedures) for those activities involving significant QEH&S risks.
- Identify all activities, products, or services that have potential QEH&S risk impact so that appropriate process, material design reviews, and risk assessments can be performed.
- Establish management control of activities, products, or services that pose significant QEH&S risk.
- Establish and support with appropriate resources risk assessment and or process enhancement tools. including employee participation.

Management must evaluate the risks and needs of external and internal customers. Evaluations should include each critical at risk task to determine its impact on the organization's collective QEH&S processes. Process flowcharts and fault tree analysis can be useful tools to help create and communicate the process risk picture, including its known or potential points of nonconformance with sub-tasks. As I mentioned above, much can and should be accomplished by using qualitative risk assessment tools as the first sieve in the facility or process risk assessment process.

Risk assessments can be very complex quantitative assessments or, oftentimes, the more applicable qualitative assessment. The purpose of risk assessments is to identify resource limitations or faults in the process design. This step allows us to develop control action plans, including changes to or the development of standard operating procedures to either eliminate or appropriately control the risks.

Organizations must have defined targets and must constantly measure or monitor their performance, including determining the multiple root causes of nonconformance or why goals were not achieved. Managers must also learn that rarely are there zero risks and that you can reach the point of diminishing returns in both the risk assessment and risk control process. Qualitative risk assessments involve employees and other stakeholders in the identification of process risk, the ranking or prioritization of the risk, and prioritizing the various control options that will be considered. Qualitative risk assessments systematically evaluate processes to identify steps in the chain of resource inputs that may cause a process nonconformance, an accident, an environmental release, or a failure to produce products/services in conformance with requirements.

Risk is a combination of event consequences, and the probability of the event or process nonconformance. Controls are developed and implemented into work processes though the management system, as illustrated in the model in Figure 11. Therefore, all losses are a result of failures of the management system, which will become systemic if their multiple root causes are not identified and if specific measurable actions with reasonable target dates (SMART) for corrective/preventive actions are not determined and initiated to enhance the process.

The goal of the group-facilitated risk assessment process is to identify the process QEH&S risks in the management system, process, or procedures. When the risks are identified, the group then ranks them for both management and control purposes. Prioritizing of the risks is based on severity, consequences, and probable frequency of the potential process failure. Existing controls and process designs are then reviewed to determine what can be done to either eliminate or minimize the risk. This logical and integrated approach to process

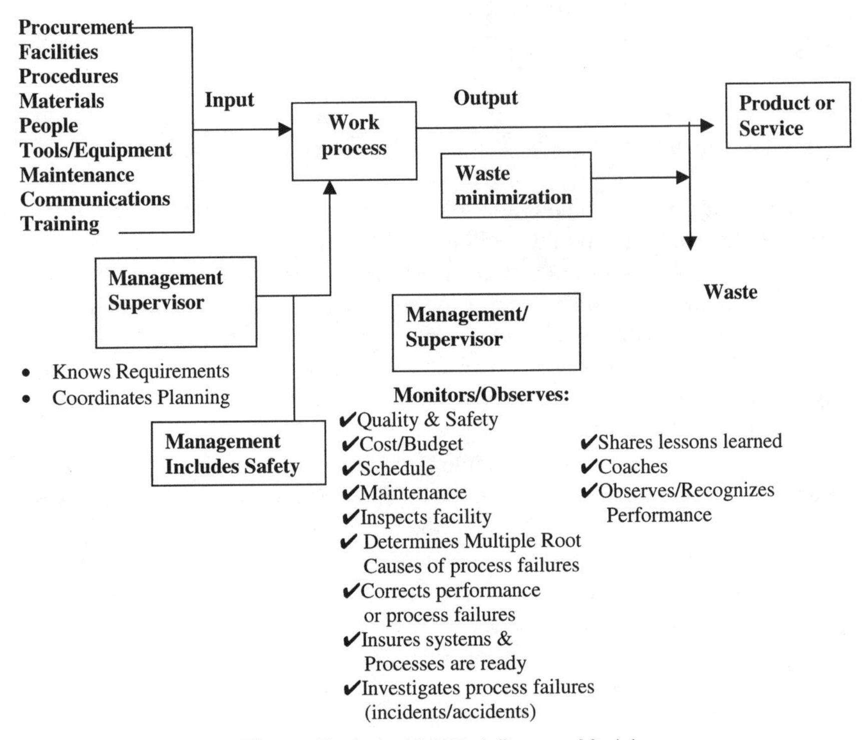

Figure 11 Industrial Work Process Model.

enhancement improves quality while reducing the risk and associated costs of quality or environmental health and safety losses.

Procedures must be evaluated periodically as part of the process management review and continual improvement process. Facilitated group process risk assessments or employee participation on process enhancement teams (PETs) is critical for process improvement and risk reduction. It allows for incremental and innovative process changes. These smaller incremental changes are more digestible than trying to revolutionize process improvement overnight. The PETs can accomplish these objectives.

- Identify risk of nonconformance and rank order the risk, its causes, and its potential control alternatives.
- Determine if the process has potential or known failure points.
- Define and improve processes or procedure design.
- Develop ways to eliminate or reduce waste, rework, injuries, or other liabilities.
- Determine how the process can be done faster, better, and cheaper.
- Establish a plan of action and milestones for process improvement SMART goals.
- Identify benefits and risk of changes, including costs, return on investment, and payback period to implement change or establish controls.

Risk assessment is the process used to determine the level of relative risk of process failure, rework (nonconformances, including environmental risks), injury, or illness associated with each identified hazard or process risk. The risk ranking/prioritization controls uses the same ranking process. Process definition and identification of procedural requirements are defined to evaluate risk, that is, the specific sequence of task steps, including consequences of the exposure to the potential risk and the probability or likelihood of it occurring. Existing controls should be identified and evaluated when a risk exposure is assessed. It is also a good practice to determine two risk rankings for processes, the pure risk (no controls) and the controlled risk. Knowing these two risk rankings will be very helpful in planning and budgeting for control resources for the organization's remedial actions and the continuous improvement process.

Organizations should determine the acceptable level of risk by reference to customer requirements, product or service specifications, EH&S legislation or regulations, standards, and codes of practice or industry best practices. Risk assessments should consider the likelihood of a nonconformance or EH&S risk and the potential severity or magnitude of the process failure.

COST-BENEFIT ANALYSIS OF RISK REDUCTION MEASURES

If appropriate, the organization should consider the present/future value impact of risk controls by using defined assumptions for internal rates of return, depreciation, risk reduction, and inflation rates.

Capital investment is made with the objective of obtaining sufficient economic/business returns to justify the outlay of resources. QEH&S managers should work with their financial managers to establish project evaluation and documentation tools to accurately measure the total cost of process nonconformance (direct and indirect costs). In addition a financial process for evaluating the costs and estimated loss prevention/risk management benefit expected from several control alternatives under consideration must be facilitated. Some relatively simple-to-use tools from Liberty Mutual or the ORC can be used for performing this type of financial assessment for EH&S control alternatives.

With whatever method used for the cost–benefit analysis, it should take into consideration the trade-off of current cash outflow versus future cash inflow or expected savings. The company's IRR or cost of capital (often estimated at 10%) should be used to discount the future cash flow or expected savings. The payback period, without consideration of discounted cash flow, is inadequate. You must consider the future value-benefits and costs. In the end you will need the following to perform your cost–benefit analysis.

- Company's internal rate of return (IRR) or cost of capital.
- Estimated inflation rate for the life cycle period of the proposed controls.
- Expected cost/annual savings generated by the proposed control.
- Depreciation cost of the proposed control alternative (as applicable).
- Expected economic/operating life of the control (measured in years).
- Cost of the proposed control, including maintenance and operating costs.

To evaluate the benefits attractiveness of the proposed controls, we will be looking at three elements: investment, operating benefits, and expected time period or life of the control. We will want to take into consideration the time value of our cash flow (net present value–future value), payback period, and rate of return on the investment alternatives.

Future Values (Compounded Values)

The future value of the investment is compounded annually over the life of the control (cash-flow period) by the company's internal rate of return or cost of capital. For example, the value of $1 compounded at 10% over a four-year period is $1.46. This calculation would be the expected benefit per dollar of investment with an IRR of 10% during a four-year period.

Present Value or Discounted Value

This present value is the mathematical inverse of the compounded value. It describes the value of the $1 paid back after four years. In this case, each dollar invested would be worth only $0.68 in today's dollars (present value–discounted value).

When performing cost–benefit analysis the finance department will be looking at the value of cash outlays made today and their value over time as the benefits

TABLE 10. Amortization of $100,000 EH&S Control Investment @ IRR of 10%

Years	Beginning Balance	Simple 10% Earnings on Investments	Depreciation	Annual Savings or Loss Control Benefit	Ending Balance
1	$100,000	$10,000	$25,000	$50,000	$85,000
2	$85,000	$8,500	$25,000	$50,000	$68,500
3	$68,500	$6,850	$25,000	$50,000	$50,350
4	$50,350	$5,035	$25,000	$50,000	$30,385
5	$30,385	$3,039	$25,000	$50,000	$8,424
6	$8,424	$842	0	$50,000	$40,734

of the investment are realized. In doing these calculations the financial advisor will also take into consideration the tax benefits of the capital investment and depreciation. A simplified example above should help to clarify these concepts and their importance.

If you have a $50,000 annual benefit from the control alternative, a simple payback period would be two years for a $100,000 investment. However, when you take into consideration taxes, depreciation of the control, and the income deferred by the company investing in the control (the time value of the investment: present–future value considerations) causes the payback period to be significantly longer and the rate of return on the investment less. As illustrated below, the actual payback period will be more that five years versus the two-year estimate made when we did not consider depreciation and the time value of cash resources. In actual investment discussions you would also need to consider tax implications associated with the investment and anticipated changes in pre-tax revenues. Table 10 provides an example of a cost–benefit analysis for a $100,000 investment with an expected $50,000 per year control benefit. As Table 10 illustrates, the payback period is significantly longer than two years when you consider the IRR on the investment and the depreciation of the control investment.

ALTERNATIVE RISK MANAGEMENT STRATEGIES

Two Aspects:

- Risk management/QEH&S — Risk Assessments
- Alternative risk transfer and finance of risk

Integrated Risk Management (Evaluation and Control of All Operational–Business Risks)

Evaluate operational risk by using quantitative or qualitative risk assessment tools and systems safety, process quality or risk assessment tools such as fish bone process flow diagrams, Haz-Ops, etc.

Controls must be focused on the elimination, reduction, or prevention of QEH&S risks. Risk assessment should also include costs or reasonably documented estimates of the costs of nonconformance, and the cost–expected benefit, return on investment, and payback period for various control options.

Categories of Business Risk

Operational

- QEH&S
- Property/facility
- General shipping and transportation/distribution
- Purchasing and inventory management
- Process control
- Maintenance
- Contractual
- Design or contract errors and omissions
- Information systems and communications
- Natural hazards (floods, earthquakes, etc.)

People—Employees, Partners, Contractors

- Business interruption
- Political
- Regulatory and legislative changes
- Financial (e.g., changes in interest rates, stock value, currency, taxes, and global transfer gaps, or inconsistencies)
- Security, property, facility, cargo, personnel, and senior managers
- Directors and officers

Strategic: mergers, acquisitions, joint ventures and alliances, resource allocation, business planning

Competition: Market

Distribution: security, environmental, distributor dependence

Emergency Preparedness and Contingency Planning

Risk Assessment Steps

1. Risk Identification
2. Measurement of loss frequency and severity (include statistical evaluation)
3. Assessment of existing potential risk and loss-control effectiveness
4. Risk map

Risk assessment must include business and financial exposures in addition to operational risks. When considering process risks, it doesn't make sense to

Figure 12 Risk Assessment.

exclude any risks that could threaten the viability of the organization. Risk management is playing a very different role today.

Years ago many risk managers were considered to be in-house insurance brokers. Today risk managers can and should play a broader role in helping organizations evaluate all forms of risk. Integrated risk management merges the tools of systems safety and financial controls. As an EH&S professional, I believe that this change will ultimately mean that businesses will realize much greater control over EH&S risks while increasing productivity and corporate profitability.

The problem with fragmented risk management is a tendency to break down the work process and restructure it so that other categories of risk are overlooked and their existing controls are compromised, thus potentially increasing costs or making the process less efficient. An integrated systems safety approach to risk management requires organizations to systematically identify, evaluate, and prioritize all risks. Once this task is accomplished, management can develop plans that appropriately allocate resources to control losses or potential risks. Plans could also include the development of operations contingency plans and emergency responsive plans.

Because risks will be better defined financially with this type of approach, it will be easier for QEH&S professionals to compete with finite resources allocated to other operational requests for funding. In addition, process improvement will be significantly enhanced by the synergy created by working with other operations and process experts such as risk manager, quality manager, and environmental health and safety manger.

RISK FINANCING

Issues to Consider

- IRR.
- Depreciation.
- Discounted cash flow or present/future value of control options, including design, installation, procedure development, and training costs.
- Payback period of return on investment.
- Other costs such as maintenance, installation and testing, training, impact on other processes such as scheduled delivery, quality, etc. should also be considered when doing cost–benefit analysis of control options.
- Expected benefits of controls or process change expressed in the financial terms listed above and/or cost per unit of production expected reductions in loss frequency and severity.

It is the understanding and balancing of these issues, especially with regards to senior management communications, that allows managers to more effectively communicate risk and proposed control options with operations, finance, EH&S, and quality managers. Equally important is to understand and be able to manage the financial consequences of realized losses. The failure to identify and appropriately manage risks has caused many companies to go out of business. To not conduct operations and process risk assessments is to accept them blindly.

Risk Finance (Integrated Risk Transfer Insurance and Alternative Financing)

Until recently, companies managed risk by purchasing insurance (risk transfer). As the cost to insure increased, larger companies pushed back and began asking their risk management team to design financial plans to manage loss in ways that produced the following results.

- Improved balance sheets
- Controlled fluctuation in profits
- Tax advantages
- Improved cash management
- Reduced administrative and third-party costs

ALTERNATIVE RISK FINANCING

- **Unfunded retention:** Pay for losses out of cash flow (like your deductible on your private insurance, that layer of risk you accept).
- **Post-loss financing:** Use lines of credit or advances on finite insured plans.
- **Pre-loss financing:** Underwriters use loss severity and frequency history to estimate potential losses. Other alternatives include self-insuring by establishing a trust fund or insurance captive. Establishing a captive creates some tax considerations, and it allows for various financing plans, including re-insuring a portion of the risk with an insurer.
- **Traditional insurance**.
- **Integrated risk financing programs** include integrated risk portfolios and multi-year/multi-line insurance coverage.
- **Indemnification** or "**Hold Harmless**" contracts to transfer risk.

To be successful, the risk manager must understand the financial condition of the company and the corporate objectives, especially those relating to quality, EH&S, risk management, and integrated process control. In addition, the risk manager should evaluate the organization's insurance needs and the various risk financing options while considering the company's risk tolerance and ability to retain loss, liquidity, and borrowing ability. Risk managers and QEH&S managers must have an understanding or process for identifying the multiple root causes of losses and an understanding of existing loss controls and a systems approach for control enhancement.

This objective ultimately leads to risk profiling or, as Microsoft's former risk manager, Scott Lange, describes it, "risk mapping." The results of the risk assessments and profiling are graphically communicated in a model, oftentimes a single-page chart. The steps in this process are listed below.

1. Risk assessment (qualitative or quantitative).
2. Statistical evaluation of loss frequency and severity: calculate mean exposure calculate the standard deviation of the data (characterize data variability).
3. Chart the risk types showing the loss severity and frequency.

The process must start with auditors and risk assessment experts who have group facilitation expertise. (see Quintan-Hall and Renner, 1940 for an in-depth discussion of group facilitation methods). These professionals will be needed to audit processes and to assist management perform both qualitative anal quantitative risk assessment.

Quantitative risk assessments should be performed only for significant potential risks. A great deal of progress can be made by using your own most valuable resources—your employees in group-facilitated qualitative risk assessments. The process requires the facilitator to be familiar with a variety of systems safety tools to consider integrated process risks or complete operational risk profiles. Some of the risk assessments may require quantitative risk assessments and/or fault tree analysis to thoroughly evaluate the level of process risk or process complexity.

An effective integrated risk management program will require involvement of the following resource and skill sets.

- Group facilitator
- QEH&S professionals/managers
- Operation and facility managers
- Security experts, transportation experts (occasionally)
- Legal experts
- Political and economics experts by region or country
- Work center supervisors and employees
- Statistical tools
- Insurance experts
- Financial management

Risk is the uncertainty of potential losses and is measured as a variable of the mean of past occurrences. Variability is defined by standard deviations, as in the examples illustrating the mean of severity and frequency associated with a loss. Two standard deviations from the mean severity are at the 95% confidence interval. Therefore, there is a 5% risk that severity associated with the loss will exceed the interval around the mean. Three standard deviations would create a 99.7% confidence interval around the mean that the loss would not occur outside the range around the mean.

In the example given below (Table 11, and Figure 13) the standard deviation is $130,000, and the range around the mean increases $260,000, going from a 95% confidence interval to a 99.7% confidence interval.

Therefore, you can see that the greater the confidence level, the broader the range of possible losses. Data quality and volume will also affect the range. The

TABLE 11. Example-Four Years of Losses

1994	$1,200,000
1995	$1,400,000
1996	$1,300,000
1997	$1,500,000

Average or Mean = M = $1, 350, 000
Variance = 1/(N − 1)[(Y1994 − M)2 + (Y1995 − M)2+ (Y1996 − M)2 + (Y1997 − M)2 = 1/(4 − 1) [(1.2M − 1.35)2 + (1.4M − 1.35)2 + (1.3M − 1.35)2+ (1.5M − 1.35)2 = 16, 666, 666, 666
Standard Deviation = Square Root of Variance = $130,000

At 68.3% Confidence Level:
1,220,000____-130,000____________**1,350,000**______+130,000____**1,480,000**
At 98.4% Confidence Level:
1,090,000____-260,000____________**1,350,000**______+260,000____**1,610,000**
At 99% Confidence Level:
960,000____*-390,000*____________**1,350,000**______*+390,000*____**1,740,000**

Figure 13 *Statistical Range of Potential Losses.*

more representative data you have, the smaller the variance tends to be, and the deviation around the mean is less. From a practical perspective it is intuitively obvious that the less fluctuation or variability in the losses or data the more confident you can be defining your risk with a narrower confidence interval.

This same statistical process is applied to quality programs to determine if deviations around the mean of expected process outcomes are random or statistically significant. If quality deviations exceed the control limits, the multiple root causes for the data fluctuation or process non-conformance should be determined. As processes are better-controlled, the variation in measured data decreases causing the confidence interval or statistical range around the data to become smaller. This finding is key to the continuous improvement process. Statistical process control (SPC) provides a tool to evaluate process deviations that previously may have been considered a random event. However, SPC identifies changes outside the upper and lower confidence intervals as significant, requiring closer management attention. The objective of further refining the process quality control is gradual improvements that will in time lead to less rework, improved QEH&S, and a narrower interval around the average or mean of current process deviations (continual improvement).

Lange's risk mapping tool is a good way to better communicate risk, including its statistical evaluation. It allows the risk manager to clearly illustrate a department or company's risk on a single sheet of paper. Naturally the graph is supported by additional data. However, the beauty of the risk map is that it allows you to clearly present a comprehensive and integrated risk picture to busy financial and operations managers, senior executives, or directors.

The risk map plots severity on the vertical axis and frequency on the horizontal axis. You can also illustrate the range of the risk severity and frequency. The risk map illustrates the impact of controls and risk transfer options, and it effectively summarizes potentially complex discussions around specific process risks. The upper left quadrant of the risk map represents high severity and low frequency losses. The upper right quadrant represents high-severity, high-frequency losses, the type that require process control intervention or avoidance; whereas the high-severity, low-frequency losses may require changes in the process and risk management by various risk transfer methods.

The lower left quadrant of the risk map represents low-frequency, low severity losses that will not require significant management attention and resources. They are a lower priority for additional process or financial controls. One way to explain the risk map and recommended controls is that as a corporate objective

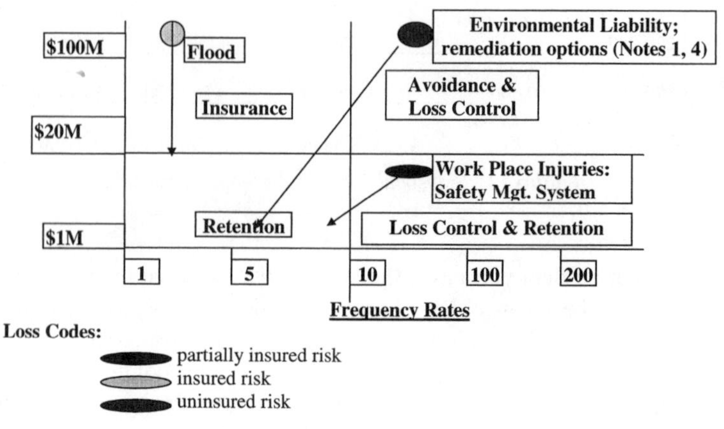

Figure 14 Sample Risk Map.

you want to reduce all operational risks to that lower left-hand quadrant of the risk map. The lower right quadrant represents high-frequency and low-severity risk, which should be controlled through process control, see Figure 14.

The line separating the upper and lower portions of the risk map is determined by management, deciding how much risk the company is willing to retain and how much to control by a combination of risk transfer and process control methods. Each of the risk circles plotted on the risk map is the point on the graph intersecting the mean frequency and severity distribution of the defined risk. As indicated earlier, you can also indicate the statistical variance and the range of potential losses on the risk map.

The sample risk map, illustrated above shows that management has decided to retain (self-insure) risks up to $20 M and to control the frequency of risk events to less than 10 per defined unit of time, dollars, facility, or production unit. The notes on environmental refer to hypothetically attached explanations of the proposed control alternatives.

With this tool, management can visualize risks and the necessary management action to control them.For example, fire loss could be $100 M, and, to control the risk, management installs sprinkler systems and fire alarms. In addition, management improves material storage methods and trains its employees. The severity of the fire risk is still high, but the probability of loss is significantly reduced, allowing affordable risk transfer with manageable loss retention. As mentioned earlier, managers will probably need two risk rankings — one that ranks the pure uncontrolled risk that continues to exist until eliminated no matter what controls are used, and the controlled risk ranking. which requires management to sustain its effectiveness.

Risk maps can show all of the corporation's risks on a single page, and more definition can be demonstrated at the facility or process level by drilling down and presenting this level of detail on attached department or process risk maps supported by severity, frequency data, and control options.

The risk map, as in the example above, can also show the effects of risk control options and information. This depiction significantly assists senior managers to determine resource utilization and risk management–process control prioritization.

The impact of controls can be visualized and balanced with supporting data documenting expected loss reductions and return on investment, taking into consideration depreciation, present values of the cash outlay against expected internal rates of return. Liberty Mutual and the ORC both have software programs to assist with these financial calculations.

Integrated programs are becoming attractive because there is often residual and unrecognized risk and additional costs associated with process management when technical mangers focus on a single aspect of organizational process risks. In some cases, technical managers working independently may complicate procedures, or even establish controls that conflict with other process expert's recommendations. Communications becomes a critical factor in ensuring that employees work in an environment that does not promote substandard performance.

In addition, from a risk transfer or insurance perspective the traditional single line of insurance coverage with multiple policies and varying retention levels and policy limits create coverage gaps and tend to be more expensive. These expenses are due to miscellaneous administrative costs and an inability of the underwriter to spread risk over a larger group. Integrated risk transfer programs also tend to be for multiple-year policies, and, by aggregating the risk, administrative and risk transfer costs are reduced. Integrated programs also eliminate the gaps created by mono-line insurance policies or policies between different carriers in different countries.

The integrated approach has a lot of potential value for multi-national corporations. Judy M. Lineman, Vice President of Insurance Risk Morgantown Fidelity Investments, estimates that savings could be as much as 30–40% (National Association of Underwriters NAU training brochure).

Integrated Risk Management Process

- Conduct audit and risk assessments.
- Review loss history, severity, and frequency.
- Perform statistical evaluation of loss history to determine loss predictability and to map risks for senior management.
- Determine retention level and need for process controls.
- Allocate reserves for retained loses and establish budgets for controlling limitations.
- Select a aggregate limit that exceeds annual losses for all covered losses (retained deductible).
- Develop layered risk transfer plan, including retained deductibles, bonds, integrated policies, and captives.

Integrated risk transfer plans include coverage for

- Property
- Business interruption
- Director's and officer's liability
- Transportation, security, and cargo risk
- Environmental
- Product liability
- Workers' compensation
- Auto
- Political
- Natural disasters

Programs can be structured with aggregated retention for the policy term or separate special occurrence deductibles. Several risk transfer alternatives provide coverage above the retention level. Other benefits of bundled–integrated risk transfer and financing alternatives include cleaning up your balance sheet releasing reserves and potential tax advantages. To fully understand the potential of this process, risk management teams should include finance, EH&S, quality, and operations managers in the risk assessment and risk mitigation process.

After the risk assessment teams have qualitatively identified potential risk, the finance and corporate risk management offices should work to statistically evaluate loss histories and potential losses identified by the group. Like loss control, the same prioritization of resources should be applied. However, understanding the financial consequences as early as possible may affect control prioritization and will be necessary to create an integrated or bundled risk transfer alternative to mono-policy plans. Figure 15 illustrates a potential risk transfer or insurance model. The amount of risk retained and held by a captive before reinsuring will affect the ultimate cost of the insurance.

Potential participants in an integrated risk management process include the following.

- Risk manager
- CEO
- Insurance broker or underwriter

Figure 15 *Layering of Integrated Risk Finance/Transfer Programs.*

- QEH&S managers
- Operations managers/supervisors

Larger companies considering integrated risk management programs will want to factor in the potential benefits of a captive or finite risk program. As mentioned earlier, captives are owned by non-insurance corporations to improve the financing of their risk transfer goals. Captives can also have multiple corporate parents, and they generally re-insure a large percentage of the loss risk.

Finite risk programs are integrated insurance plans where premiums are paid into a managed experience fund held by the insurer. This plan in effect allows the insured to pay for its losses over time with regularly scheduled payments. If loses are less than expected, the insured company shares in the profits and investment income earned on the premium paid into the fund.

Finite programs are usually multi-year policies with some aspects of traditional insurance and most often create an aggregate loss limit and defined retention for various risks joined in the aggregated plan. The higher-level risks are generally covered by a conventional stop-loss layer. The insurer assumes the timing risks if losses exceed the funds paid into the plan. However, the insured must still pay the insurer for the losses paid out in advance or loaned to them as participants in the finite plan.

Larger companies with high retention levels use finite programs. The plans can improve balance sheets and smooth out earnings by eliminating the unpredictability of losses. Companies considering finite plans should evaluate how much they currently have in reserves, the amount they self-insure or premiums paid for various plans to insurers, and the gaps they may currently have in insurance coverage. Gap analysis should be performed by looking at coverage between countries, facilities, states, and the assorted insured plans.

Finite plans can be difficult to set up and must be considered in light of tax and accounting rules that require real risk transfer underwriting with reasonable loss expectations to be tax deductible. Funds and interest earned at the end of the policy period are also considered income unless rolled over into a renewed plan. IRR on the large cash flows to the fund must also be considered.

RISK ASSESSMENT PROCESS SUMMARY

- Risk assessments should be used during operations reviews and when processes are altered.
- Risk assessments systematically evaluate potential problems.
- Risk assessments identify and evaluate controls and the need for additional controls until the risk is eliminated.
- Generally you compare non-conforming tasks to problem-free processes or process designs or procedures prior to incidents or accidents, nonconformances.

- Risk assessments identify and list potential or known process deviations.
- Risk assessments then analyze the potential deviations and risk rank them.

Risk assessments must consider risk to the following:

1. Employees
2. Contractors
3. The public, neighbors
4. Environment
5. Customers and Distributors
6. Product Quality–Consumer Expectations

Safety tends to be something that ebbs and flows with the budgets of companies and oftentimes is a reflection of general economic conditions. In other words, we too-often sacrifice safety as it becomes necessary to tighten the budget. Safety also tends to have more support for projects or risks with greater emotion; it is less understood; or it has greater visibility. Therefore, limited resources are at times allocated to controls without consideration for their overall risk ranking of the targeted problem. Corporate or facility-wide risk assessments are necessary to ensure that well-designed procedures are not compromised by inadequate training or procurement controls and that high-risk tasks are targeted first for intervention.

To the extent possible, all non-essential risk should be eliminated, whereas the remaining risks are controlled to minimize the risk potential. Therefore, the nature of the risk must be understood before controls are designed into the process. Fault tree analysis and other system safety tools such Hazard & Operability (HAZOP) or Management Oversight Risk Tree analysis (MORT) can be effectively used to better-define existing processes. Change analysis is also an effective systems safety tool that can be very helpful when investigating process nonconformances or accidents. No design should be considered flawless. It is our job to identify and minimize the risks.

MORT was developed by William Johnson in the 1970s to evaluate critical processes before they are implemented to ensure their readiness. It is also a comprehensive tool to determine the multiple root causes of process non-conformance. The MORT user's manual states:

> This report is the User's Manual for MORT, a logic diagram in the form of a work sheet that illustrates a long series of interrelated questions. MORT is a comprehensive analytical procedure that provides a disciplined method for determining the causes and contributing factors of major accidents. Alternatively, it serves as a tool to evaluate the quality of existing systems for readiness and safe operability.

In reality, MORT and other fault tree or systems safety tools can be used to evaluate the general quality and process risks associated with a system because the related process risks could be determined by using the same tool. The goal of the system is to prevent injury, rework, and waste of any kind, including environmental or hazardous waste releases.

Over the past 30 years we have seen stricter EH&S regulation and enforcement, including criminal charges, murder convictions, and, more recently, indictments for murder. At the same time, the public and consumers have become much less tolerant of poor QEH&S management. Consumers are demonstrating their views by boycotting products and corporations. They have also won significant liability law suits, including the recent $4.9 B award against GM for knowingly selling vehicles with unsafe gasoline tanks.

The QEH&S life cycle (product) risk assessment must scrutinize the risks from cradle to grave or from product design and production planning to shipping and consumer use, all the way through to disposal, reuse, or recycling. Identification of risk could result in product, process, production, use, or packaging redesign to mitigate the risk to the lowest level possible. Unmitigated risk, including potential disposal liability, should be factored into the product's cost and the company's continuing efforts to eliminate the risk/liability.

Management must understand the multiple root causes of process nonconformance and QEH&S risk. Although regulations can and should be a motivator, in the long run a corporate focus and process to identify and control risks will be much more effective in improving QEH&S. It will also better-ensure regulatory compliance. EH&S regulations are vast, and they vary from jurisdiction to jurisdiction. A tremendous administrative and legal effort can be utilized trying to stay ahead of the regulators. Companies must comply with the law, but the basket of regulations that you must be concerned with can be significantly reduced and better-managed, by identifying potential risks and then either eliminating or reducing them, the potential regulatory risk will automatically go a long way towards conformance.

In some ways, the above statement is like the old question of which came first, the chicken or the egg? Evidence of the tremendous waste in capital and resources generated by taking a regulatory or legal defense posture is clearly evident in the fact that most of the Super Fund money has been squandered between bureaucrats and corporate legal teams. If you understand the causes of risk and can clearly define your processes, including risk exposure pathways, then you will be well on the road to containing the risk and enjoying regulatory compliance. Part of this process must also include defining risk and control alternatives in financial terms so that appropriate management decisions can be made.

In addition, clearly quantifying potential financial impacts of defined risks is necessary not only to ensure accurate product costing, but it is required by Financial Accounting and Standards Board (FASB) and the Securities and Exchange Commission (SEC) in the United States, and more recently by the Turnbull regulations in the United Kingdom. Part of product expense may be to contribute to financial reserves established to offset contingent EH&S or product liabilities.

QEH&S planning must include a review of product life cycles to determine the following.

- How can the product or its components that have reached the end of their life cycle be positioned for reuse or recycling versus disposal?
- How can we reduce the potential exposure to employees, consumers, or other stakeholders?

After developing management plans and communicating these objectives via changes to or by the establishment of policy and procedures, companies must define the processes for implementation in each department or business unit affected by these policies. Product life cycle and/or work center process enhancement teams must be established to identify QEH&S risks. Some of the tasks assigned to these teams can include the following.

1. Identifying existing process and product life cycle QEH&S risks or liabilities.
2. Estimating the financial impacts of these risks to the extent possible. If data are not available, you might be able to define best-case, worse-case, and most-likely case scenarios, including a rating regarding the probability of each scenario. Weighing these potential cost scenarios with a relative probability number 0–1 will allow you establish an estimate of the probable costs or contingency/reserves that will need to be established.
3. In the facilitated process enhancement team meeting, identify possible prioritized controls, including reasonable cost estimates for full implementation of the various control alternatives by using discounted cash flow methods to document the cost–benefits of control alternatives. Discounted cash flow projections should be calculated, considering after tax cost–benefits of recommended control alternatives including net rate of return, and the payback period for each alternative.

   Note: QEH&S process enhancement teams must be able to recognize the need for specific technical support from other departments such as finance, marketing, research, and engineering.

Document total QEH&S costs related to product/service production and QEH&S nonconformance estimates (defining assumptions, if used) and future QEH&S costs, including end-of-service-life disposition liabilities. Assign these costs as potential overhead liabilities. These, liabilities should not be pooled and, to the extent possible, they should be charged back to the work center generating liability or risk so that appropriate product costing and work center budgeting can be done.

Here are some examples of waste reduction. Ford issued design guidelines (1993) to its suppliers to improve recyclability of its vehicles and increase the (vehicle) recycled content. Nearly 80% of the content of its new cars can now be recycled. It is using many other recycled materials in its production, including

54 million recycled soft drink bottles to produce luggage racks, reinforcement panels, and door padding. It also moved aggressively to reduce solid waste and is now recycling more than 70% of the solid waste material from its plants... ATT has employed total quality management (TQM) principals and reduced paper use in its offices by more that 25% and millions of dollars in three years. The company is recycling more than 60% of its waste paper and applying the design for environment approach to manufacturing so that the design of any product will take into account the environmental impact of its entire life cycle (Epstein, 1998).

Figure 16 is provided to assist you with prioritizing control actions and follow-up. Figures 17 and 18 illustrate the process or packaging necessary for life cycle risk analysis. At the end of this chapter I have also included some forms that can be used to break down processes to better understand their inherent risks.

- Use life-cycle analysis to better understand and estimate full life cycle cost accounting: product design, production process costs, storage, distribution and shipping and ultimate post-service disposition (internal/external and direct/indirect), pricing and liability, and improved management decisions.

Procedures should be established and maintained to identify hazards and assess risks related to the activities, products, and services that the plans impact or when there are significant design, material, or process changes that can impact quality and/or EHS controls. The specific application of hazard identification or risk assessment and control procedures must be part of the on-going planning process. Job–task analysis or risk assessments should be performed on critical work activities to ensure that safe procedures, equipment, tools, and facilities support the productive work process.

Organizations should establish documented procedures to identify, access, and understand all legal and other requirements relevant to facility activities, products, or services. To facilitate keeping track of legal requirements, an organization can establish a list of all laws and regulations pertaining to its activities, products, or services.

Relevant information on legal and other requirements must be communicated to employees. Organizations should be able to demonstrate objective evidence of standard conformance. Regulatory compliance could include these elements.

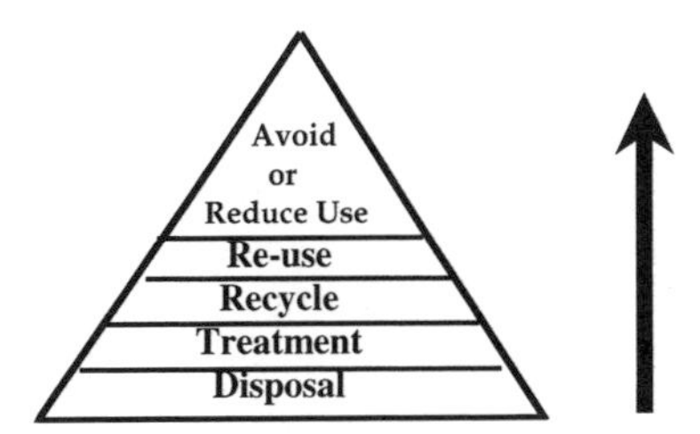

Figure 16 *Preferred Waste Management Process Hierarchy.*

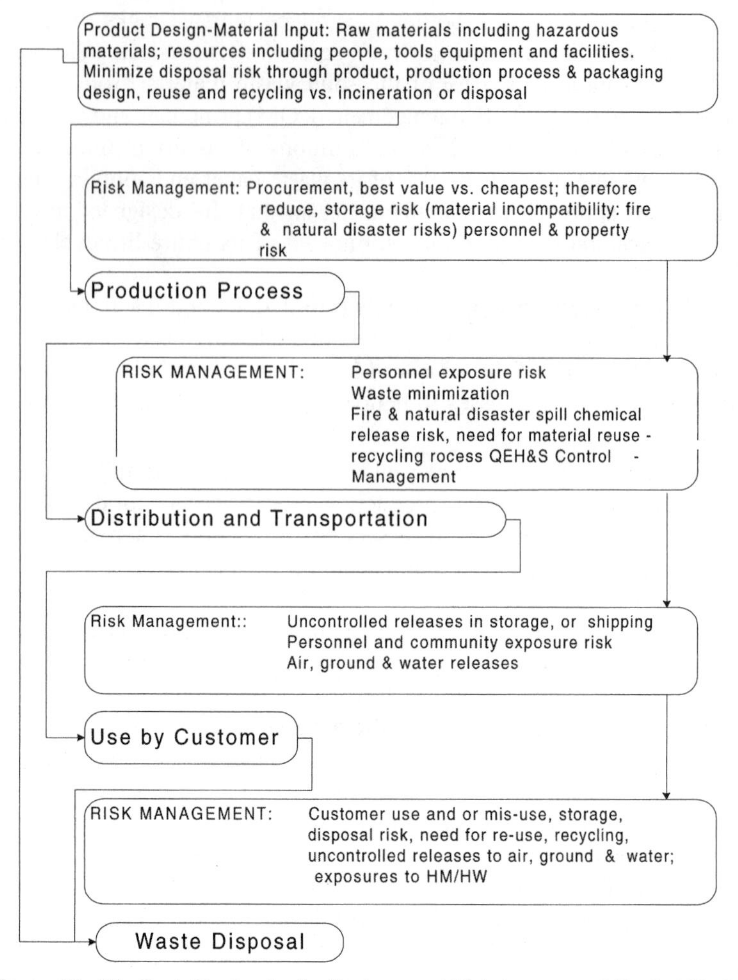

Figure 17 *Life Cycle/Design for the Environment Risk Assessment Control Model.*

a. Requirements specific to the activity (e.g., confined space regulations), emission point control.
b. Controls specific to the organization's products, services, or industry.
c. General QEH&S regulations, including authorizations, licenses and permits.

Objectives and targets can apply broadly across an organization or more narrowly to site-specific or individual activities. Objectives and targets should be defined periodically by senior management in collaboration with work center managers or supervisors and should be reviewed and revised taking into consideration the views of the interested stakeholders.

The initial review provides the user with information concerning the current status of the QEH&S management system. This information can then be used to

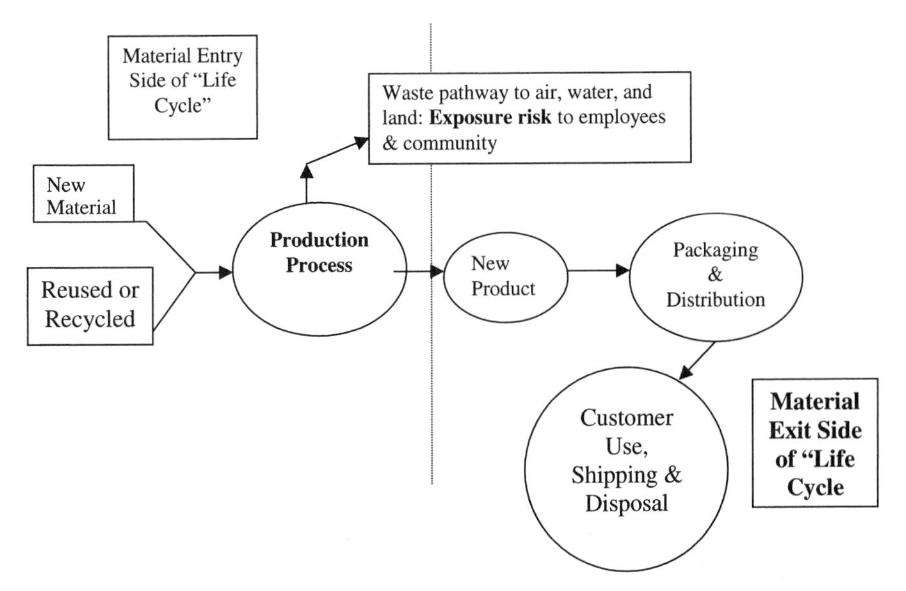

Figure 18 *HM/HW "Life Cycle"/Production Risk & Minimization Model.*

identify those work areas, practices, or activities within the organization where performance is weak. Objectives and targets, consistent with the organization's policy, should then be set based on improving QEH&S performance.

When the objectives and targets are set, the organization should establish measurable performance indicators. These indicators can be used as the basis for a performance evaluation and can provide information on process management and operational control systems. Performance measures are used as benchmarks in the continuous improvement effort.

Most companies would like a clearly defined, systematic risk assessment process that is not restricted to any single aspect of the business. In addition they want practical, well-thought-out recommendations to mitigate risk on a prioritized basis. Controls or recommendations must be country-, operations-, and site-specific.

Risk Assessment and Mitigation Process

Facility managers, supervisors, and employees are oftentimes the best source of information with regards to risk identification and risk control options. Therefore, facilitated group risks assessments that include process experts, technical experts as warranted, and facility managers, supervisors, and employees should be considered as participants in the process.

Group facilitation and the use of nominal group techniques is an art that becomes more effective with experience. The more you do it the better you and the group become. Group facilitation is important and operates on the premise that the employees and supervisors working at the task level are for the most part the true experts. They oftentimes know what is needed or have ideas of how to reduce processes risks and improve productivity and quality but are frustrated by organizational cultures that really don't hear them. Worse yet, too often they are seen as complainers and in some cases their in incentive to to tackle process problems is diminished.

As indicated earlier, qualitative risk assessments can be very valuable tools and in most instances will provide a complete understanding of process risks at a fraction of the cost for quantitative risk assessments.

Step 1: Plan the Assessment

Familiarize the PET with the organization records, loss history, and applicable benchmark data if available. Management must agree to the goals/objectives and the schedule for the risk assessment process, including initial documentation of controls and mitigation strategies.

When planning time for group-facilitated risk assessments assume that at least 50% of the group's time will be devoted to fact analysis in stage 1. Developing recommended corrective actions, correction action plans, cost–benefit analysis and follow-up will take the remainder of the time. As you can see the most difficult part of the process is adequately defining the process, analyzing the facts, and history of known or potential nonconformances.

A key role of the facilitator is to help the group define real problems and to avoid analysis paralysis. Once this has been accomplished, the facilitator must ensure that all members of the group participate in the process of brain-storming and generating ideas, evaluating ideas, and coming to consensus on recommended corrective/preventive actions. To be effective the facilitator must remain neutral. Therefore, it may be difficult for the work center supervisor to facilitate the group. The facilitator may be a member of the risk assessment team, but a third-party facilitator will significantly increase the likelihood of a successful outcome for the group.

Another key role in facilitated risk assessments is that of the secretary or recorder to help document and manage the process. The recorder can also help establish neutrality in the process by not being stuck in the group's conversation and on occasions can help interpret comments that may not be heard or understood correctly by some of the group members. The secretary recorder also helps the facilitator to review the group's activities and to provide feedback. The recorder functions primarily as the group's memory.

The members of the group must be committed to being actively involved in the process, otherwise they must be replaced. The key issue in front of the team is getting consensus on the definition of the problem before moving to problem resolution. One of the challenges for the group is being respectful of each other's ideas and views of the problem and being open to new ideas. Sometimes solutions arise from the integration of ideas, or a resolution may come about as the result of a far-out suggestion triggering another more realistic idea.

Sometimes technical experts are needed during group-facilitated risk assessments. These experts can assist with issues concerning a problem or they an help design recommended corrective/preventive actions. Here are some key considerations for a group-facilitating process.

- Define the problem, its issues, and its history.
- Define success.
- Determine what records, experts, tools or support will be needed and when.
- Assess the commitment of the team. Are the team members the right ones to work on the problem? If not, who would they be?
- Determine if team members delegates or representatives? (You really don't want representatives. The team needs delegates who have decision-making authority.)
- Determine who owns the problem, and who will be responsible for implementing the corrective/preventive actions.
- Assess the scope of risk and the worst-case scenario.
- Agree to rank problems and work only on highly ranked issues first.
- Don't jump to solutions too early. You must have consensus on the problem definition first!
- Separate fact from assumptions, and state the problem as a question (How, When, Who, When, What).
- Ensure that the group understands or determines the multiple root causes of the problem.
- Recognize what else is out there, who has similar problems, and what was done.
- Use nominal group techniques, voting, and ranking among each group member and then as a group to get to consensus on recommendation, especially if more than four approaches are being considered.

Step 2: Identify the Risks

Identify the operational risks in the various departments and in processes controlled by or associated with the department. This task can be accomplished by loss record reviews, customer complaints, rework, environmental permits, accident records interviews, job–task analysis or observations of critical tasks, and possibly audits of some business or work processes. These referenced data sources should be specifically referred to, attached, or made available to the risk assessment team, and documented in the completed risk assessment report.

In addition the department, facility, or process risk assessment team should evaluate the lists of identified risks and define the process steps, including group brain-storming of the critical process. The risks can be further evaluated by using flow charting or other systems safety checklist processes (HAZ-OPs, etc.) to define and prioritize as many operational risk and potential process deviations as possible.

The team must achieve consensus on the facts and a definition of the problem. If this consensus cannot be achieved there is little hope for developing effective corrective/preventive actions.

Decisions based on consensus will have a much greater chance of surviving and maintaining support for recommended actions. It will take longer to achieve consensus but penny-wise and pound-foolish decisions in this type of activity will tend to haunt you and could easily result in a wasted effort. The group should agree to a majority vote only as a last resort, recognizing that the process is at risk if key participants are not part of the consensus.

The first step in facilitating a group for qualitative risk assessments is to understand the process. You must begin by diagramming and defining the process as it is expected or should be. At the same time, you must document those steps in the process that have a known history or potential for nonconformance (fact analysis). After the risks are inventoried, they should be evaluated and prioritized. The group can then begin the process of brain-storming to generate corrective/preventive action ideas or modifications to the process that will reduce costs and make it more efficient. Like the inventory of risks, ideas will need to be ranked in an attempt to gain consensus for the recommendations.

Step 3: Identify the People and/or Processes at Risk

This step is necessary to identify who will be exposed to the identified risks. Consider all the people or the processes that are exposed to risks and could potentially generate risk or operational losses. With regards to QEH&S risks, it is also necessary to consider any specially vulnerable products, services, customers or groups (e.g., people with language difficulties, young or handicapped people, temporary workers, etc.). Furthermore, consider how well-trained and informed people are, particularly with regard to the risks to which they are exposed. Prioritize risks by hazard rating for risk/loss control considerations and/or risk transfer. This prioritization can include the following risk factors.

- Business operations
- Financial
- Political
- Market or public relations
- Safety
- Product quality
- Environmental

Step 4: Identify the Exposure and the Working Environment in Which the Risk is Present (Also Refer to Any Incident History)

Identify how exposure to the operational risk is generated and under what conditions the loss can be realized. Describe the consequences of exposures by considering working conditions and by using process-flow diagrams of the work and its potential deviations from the required or assumed process or procedure. In addition, evaluate the potential severity of the nonconformance or EH&S risk.

Consider the likelihood of risk creating injury/damage, rework, process inefficiency, and financial loss of any kind

Past — Incident/History/Experience
Present — Value/Judgement/Current Standards
Future — Probability/Value Judgement

Severity and frequency of the predicted outcomes are necessary to estimate the degree of risk and to forecast the potential financial loss. These estimates are based on loss history or documented assumptions used to construct a potential loss scenario and to report with documented frequency and severity the various process failures.

Step 5: Identify Control Measures and Analyze the Controlled Process Risk

Evaluate adequacy of existing risk controls. The design and potential for nonconformance or deviation from process design, procedures,and existing prevention and risk mitigation controls, including finance and risk transfer, should be evaluated as part of the process risk assessment. Control procedures may be contained in company documents, operational or quality procedures, safety policy, safe working manuals, loss reports, or insurance policies. These procedures may also be observed during the site inspection or audit. Documentation should be referenced rather than repeated in the risk assessment documentation process. Evaluate risk probability and frequency of deviation or nonconformance. Estimate impact of worst-case scenario.

Confirmation that the control measures effectively comply with standards must be documented. If control procedures are written but are not observed in actual working practices, then workforce training must be considered. Recommended corrective/preventive actions should also be ranked not only on risk mitigation or effectiveness, but also on return on investment, pay-back period, and expected reduction in losses. Cost–benefit analysis, even if simply outlined, will be much more powerful than requests for resources without a financial/economic justification.

When the group has completed its inventory ranking of recommendations and cost–benefit analysis, it will need to outline a plan of action and to assign responsibilities and due dates for approved actions. As part of the management review and evaluation process, all approved plans of action will need to be followed-up to ensure that they have been effectively implemented on time and within budget.

Step 6: Is Risk Acceptable?

If not, redesign process controls and reevaluate. Consider need for or adequacy of existing risk transfer (retention, balance sheet and tax implications, captive, and layering options, reinsurance), financial plan, and risk management program.

TABLE 12. Risk Perception Factors

Risk Acceptability	
Underestimated	*Overestimated*
Perceived vs. Actual risk	Voluntary vs. Involuntary
Known	Unknown
Little Media	Much Media
Me in Control	You in Control
Common	Uncommon
Mundane	Dramatic
Not Controversial	Controversial

Risk-hazard or consequence X-probability

Risk Assessment Categories

As illustrated above, the public as a whole does not understand risk, especially risks outside its direct control, such as those in the workplace. Risks that are not directly controlled or understood tend to generate a very low tolerance from individuals. Outside the workplace individuals will smoke, participate in high-risk sporting activities, or live a life style that generates significant risk to their health and safety—but these factors are directly controlled or understood by them. Better tools are needed to help communicate risk so that intelligent, informed decisions are made with regards to risk identification and risk ranking, especially in the workplace. Therefore, tools such as risk mapping become important when trying to explain risk or risk management options.

Most people understand the risk associated with asbestos exposure but do not understand the tremendous waste in limited resources that was applied to asbestos removal in public buildings. The potential for exposure, and thus the level of exposure for the most part, was ignored until billions of dollars had been spent in many cases on asbestos removal that could have very easily been put to better risk control in other more significant areas. This comment is not to say that asbestos is not a risk. Asbestos is a known carcinogen; however, the level of potential exposure risk in buildings may not have been appreciably reduced by full-scale removal. In some cases, exposures may have been generated to the asbestos or by a risk generated with chemicals used to strip out mastic and other bound asbestos products that was a greater risk than managing the asbestos in place.

Workers still continue to smoke, willingly accepting that risk, until they become sick, at which point they look for someone to blame for their own decisions. This observation does not overlook the attempts by the tobacco companies to sell tobacco to children, to increase the addictive strength of cigarettes, and to look the public in the eye and deny the risks associated with smoking. *This behavior is simply not acceptable!*

However, as a former cigarette smoker and occasional smoker of cigars, I accept the consequences of my own actions. I also ski and occasionally will ski advanced black diamond runs with my son; but again, I make controlled

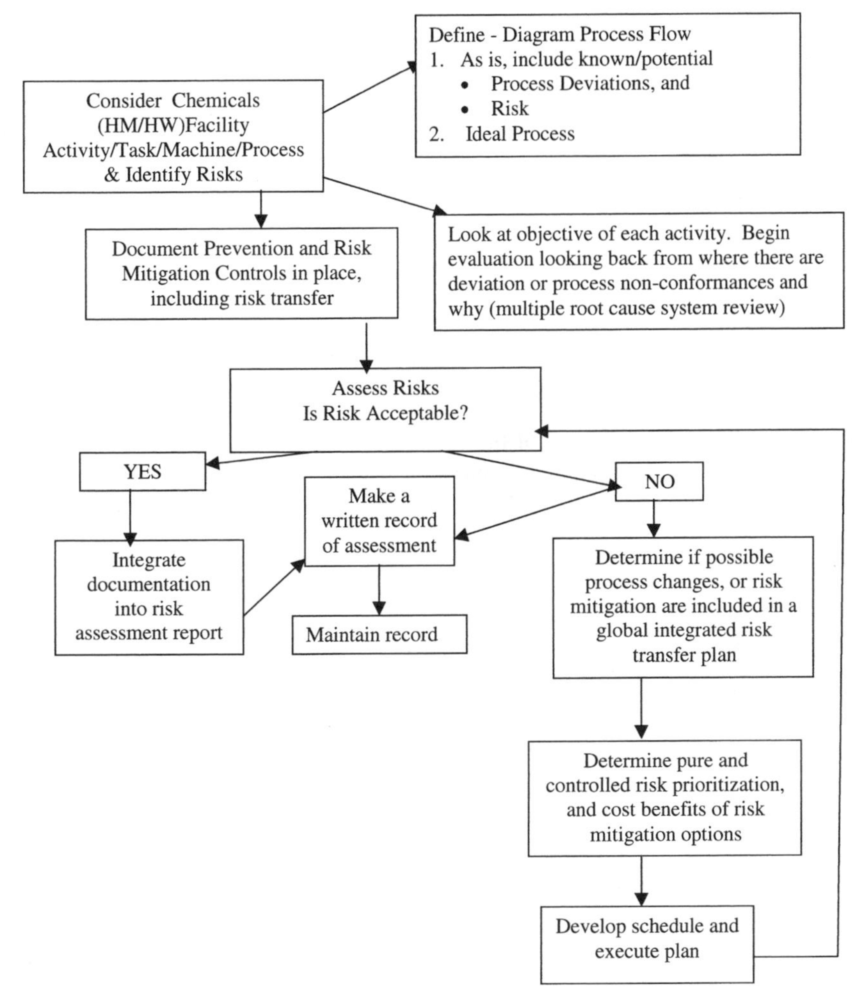

Figure 19 *Risk Assessment Process.*

conscious decisions about the level of personal risk. In both of these examples I, like most people, will tend to accept greater risk if it is outside my work and, more importantly, under my control.

During 1992 and 1993, several widely publicized murders of tourists in Florida caused millions of potential visitors to avoid the state. In fact, of the 40 million tourists who had visited Florida, 22 were murdered, which is a rate that was approximately one-third of the average number of murders in American cities. Being a tourist in Florida, in short, was a great deal safer than being a resident in most places where the tourists reside. (Mark Andrew, Liberty Risk Services, Quest 99 paper)

Figure 20 is a process model to help you understand and communicate a risk assessment process. Note that the first step is defining the process. This step

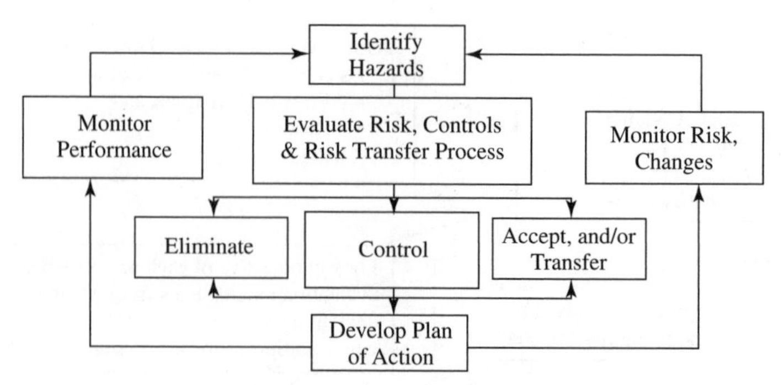

Figure 20 *Integrated Risk Management Process & Project Plan.*

should consume 50% or more of the risk assessment effort. Without a thorough, well-thought-out definition of the process, it will be difficult or impossible to determine effective QEH&S process changes or controls. There are various ways to assign risk ranking to process risks when performing qualitative risk assessments, such as the ranking categorization listed below.

Risk Ranking

High: A risk that could result in a person receiving a fatal or serious injury, a major health risk or a risk in which several people could receive lesser injuries, a serious fire, or significant risk to market or financial loss (e.g., political risk, theft, or accidental release of a customer's hazardous materials during shipping).

Medium: A risk, which could lead to a serous incident if not controlled, loss of a client, or significant damage to a product or service.

Low: A risk of minor injury or delay in the work process.

Trivial: Minor issues, which do not require immediate allocation of control resources.

Risk Assessment Review and Follow-Up. The risk assessment team should review assessments after a realistic interval (typically 12 months) or if organization or other processes substantially change to determine on-going conformance and/or the effective implementation of targeted action plans. More importantly, management must document and follow-up on recommended actions if the process is expected to effectively mitigate both process and financial risks.

Risk assessments must include business and financial exposures in addition to operational risks. When considering process, risk it doesn't make sense to not consider all risks that could threaten the viability of the organization. Equally important is to understand and manage the financial consequences of realized losses. The failure to identify and appropriately manage risks has caused many companies to go out of business. Alternative risk financing tailored to operational needs should be evaluated, including the following.

TASK:

Plant Location: **Date:** **Risk Assessor/Team Facilitator** **Department:**

Step in Operation Task or Event	Potential Incident/ Accident; Unwanted Energy	Caused By:	Step or Task Changed	Probability	Consequence	Risk Rank	Current Controls & (Used, Not Used, Failed, Impractical)	Recommended Barrier Controls & Mgt. System Changes

Basic Risk Factors: Design, Communications, Supervision Training, Procedures, Tools & Equipment, Materials, Maintenance, Incompatible Goals

Risk Ranking: High Risk= (1) A risk which could result in serious injury or death, major health risk, or where several could receive lesser injuries, a serious fire threat, significant environmental liability/release, significant threat to market or financial loss. **Moderate Risk=(2)** A risk that could cause a serious incident/accident, loss of client, or significant damage to product or work in progress. **Low Risk=(3)** A minor injury or delay in the work process (Keep in mind Probability, see attached Risk Matrix)

Figure 21 *Multiple Root Cause Analysis Risk Assessment Form.*

Occupation Or Title | Department | Inventory Date | Inventoried By: | Reviewed By:

TASKS OR ACTIVITIES	PROCESS RISK EXPOSURES EVALUATION	Risk Ranking	CONTROL EVALUATION (S=NS)			
List all critical tasks or activities normally done or that might be done by a person in this work center	Consider Process Risks: Quality/Rework ,EHS, Damage, Fire, Etc., **Consider Process Inputs/Resources/Basic Risk Factors**: Design, Communications, Supervision Training, Procedures, Tools & Equipment, Materials, Maintenance, Incompatible Goals; Have there been process changes; is there potential for deviation	H-M-L	Q	E	H	S

Risk Ranking: High Risk= (1) A risk which could result in serious injury or death, major health risk, or where several could receive lesser injuries, a serious fire threat, significant environmental liability/release, significant threat to market or financial loss. **Moderate Risk=(2)** A risk that could cause a serious incident/accident, loss of client, or significant damage to product or work in progress. **Low Risk=(3)** A minor injury or delay in the work process (Keep in mind Probability, see attached Risk Matrix)

Figure 22a *Significant Processes Inventory Worksheet.*

TASK:

Process: To more effectively identify the Basic Risk Factors & Multiple Root Causes of Process Nonconformance draft a simple process flow diagram depicting how the process was expected to happen and how it was actually done. For each step in the process that was deviated from or changed use the form below to accurately determine why it occurred and what needs to done to prevent a reoccurrence and to "Share Lessons Learned"

Plant Location: **Date:** **Risk Assessor/Team Facilitator** **Department:**

Work –Process Basic Risk Factors	Accident Free Process	Non-conforming or Accident Process	Difference and Why	Impact-Contribution to Accident or Potential Contribution to Future Process Failures
Energy Barriers				
Design				
Communications				
Supervision				
Training				
Procedures				
Tools & Equipment				
Materials				
Maintenance				
Incompatible Goals				
Process Flow				

Figure 22b *Change Analysis Evaluating Barriers & Basic Process Risk Factors.*

- Tax and balance sheet implications, including releasing reserves for recognized risks, and loss portfolio transfers.
- Different structuring of the risk-finance plan, including integrating and changing aggregated retentions and risk transfer.
- Global gaps in coverage.
- Present/future values of cash flow and impact to earnings.
- Stabilization of losses and reported income.

All risks should be identified and mapped on department or facility risk map to summarize specific operational, risks including severity and frequency of losses and the financial impact of proposed controls. The impact of controls can be visualized and balanced with supporting data documenting expected loss reductions and return on investment.

Process Summary

- Conduct audits, and risk assessments.
- Review loss history, severity, and frequency.
- Perform statistical evaluation of loss history to consider loss predictability and to map risk for senior management.
- Determine aggregate retention levels for an integrated plan.
- Prioritize risk control actions and begin risk mitigation process.
- Allocate reserves for retained losses and establish budgets for control implementation.
- Determine aggregate limit that exceeds annual losses for all covered losses.
- Design layered risk transfer plan, including the benefits of a captive or finite risk program.

Action Plan

- Establish qualitative risk assessment process of critical operations for QEH&S risks.
- Review loss history: customer complaints, rework, environmental liability, employee injuries; include cost, average severity, and frequency of the events.
- Develop inventory of business and market risks, develop mitigation and/or contingency plans.
- Evaluate and map organizational risks.
- Access feasibility of integrated risk transfer and alternative risk financing options.
- Rank risks and develop prioritized control/prevention strategies, including cost–benefit analysis of control options.
- Follow-up on mitigation plans.

6

Risk and Operational Control

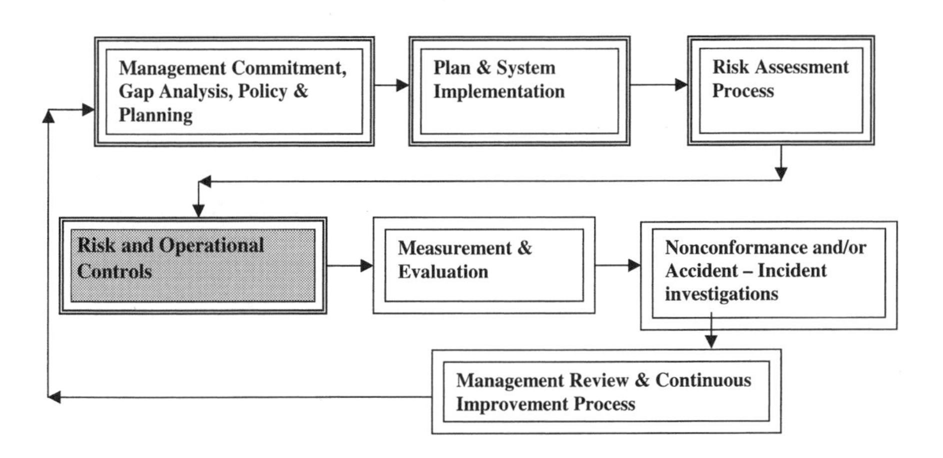

> The root causes of most safety significant events [were] found to be deficiencies in plant organization and management, the feedback of operational experience, training and qualifications, quality assurance in the maintenance and procedures, and the scope of corrective actions.
>
> —*International Atomic Energy Agency on Chernobyl (1989)*

Operational control depends on the following practices.

- Understanding work processes.
- Identification of risk.

- Process design to control the system/environment and prevent conditions that cause substandard performance.
- Having the right tools, equipment, facilities and maintenance to support these resources.
- Team work and employee involvement.
- Communications, including procedures, training, and supervision.
- Measurement, observations, management review, and focused leadership.

Organizations must have plans to implement systems, processes, and procedures for management control of work and products or services posing QEH&S risk. To start this process, companies or their facilities must identify their critical processes and perform risk assessments as discussed in Chapter 5. The purpose of the risk assessment is to better-understand the process and to prioritize QEH&S risk controls for the evaluated processes or tasks.

This chapter focuses on implementing operational controls. This process requires special attention after completing a risk assessment, prioritizing risk and various control options, and implementing corrective/preventive actions. Controls should start at the procurement process, which is the purchasing of vendor-supplied materials or the contracting for services. Controls must also be considered in the design stage and in process engineering for risk elimination or mitigation. It is also necessary to ensure the high-risk materials (hazardous materials/hazardous waste [HM/HW]) are replaced with less hazardous materials or eliminated from the process when ever possible. HW treatment and recycling are important to eliminate the long-term risks associated with waste disposal.

Ensuring that risks are mitigated to the extent possible requires a strong focus on communications, including training, supervision, and procedural controls. Inadequate communications is the single greatest cause of substandard conditions, leading to undesired behavior and work performance.

Facilities must identify all activities, products, or services that have potential quality or EH&S Risk. Documented procedures should require hazard identification and risk assessment during the design or process redesign stages (systems safety reviews). Process enhancement teams (PETs) should perform process and material design reviews or risk assessments and, as appropriate, make changes to existing design or process controls. Operational control requires an understanding of the process flow, known or potential deviations, and risk or adequacy of resources, procedures or communications, including the following.

- Purchasing
- Tools
- Equipment
- Facilities
- Materials
- Supervision and management

- Communications and training procedures (work procedures, standard operating procedures [SOPs] or process control instructions [PCIs])

After completing the targeted risk assessments, one of the first initiatives on the path to continuous improvement will be to establish and maintain documented work practices (or equivalent procedures) for those activities involving significant quality or EH&S risks. In addition, International Standards Organization and good management practice will require that a measurement or monitoring system be created to provide managers with a systems feed-back loop. It is critical that there is an ongoing review process to ensure the effectiveness of controls and to follow-up on targeted improvement actions. Management will basically get what it measures or monitors.

Part of the follow-up and management review process is to conduct periodic audits, including third-party audits. The purpose of the audits is to measure conformance to policy and procedural requirements as defined in the QEH&S manual, policy, and implementing procedures or technical requirements and to ensure that proposed process corrective/preventive actions have been effectively implemented.

Use PETs to evaluate the work center or operational risks, process design/readiness, and ensure that basic process inputs identified in Figure 6 are appropriate, designed, or intended for the job. The PET should include work center employees and supervisors, as well as production engineering, quality and environmental health and safety (QEH&S). The team must not only identify known risk, but be prepared to improve or change the process inputs to ensure QEH&S performance. The PET can be an invaluable asset for each work center. These employees probably know more about process risks and needed controls than management, and in many cases the design engineers. The qualitative risk assessment team or PET are simply vehicles to identify and control process risk, and implement procedures. They are the people "closest to the action" and who will benefit the most from effective change.

Critical qualitative process risk assessment is the most commonly used approach for hazard identification and is legislatively required in many countries. Figure 23 is a simple model for this continuous improvement process. Risk assessments predict and prioritize process risks and a proposed hierarchy of control methods from preferred (elimination) to the least desirable control methods, as illustrated below.

HAZARD RISK IDENTIFICATION

Risks should be evaluated in all aspects of the production or service delivery process, including design or purchasing/contracting, production, delivery, and maintenance.

Tools/Resources Used to Assist in the Identification of Hazards

Consultation. People who may have experience in aspects of the job may help identify potential hazardous activities or process risks. The identification of some

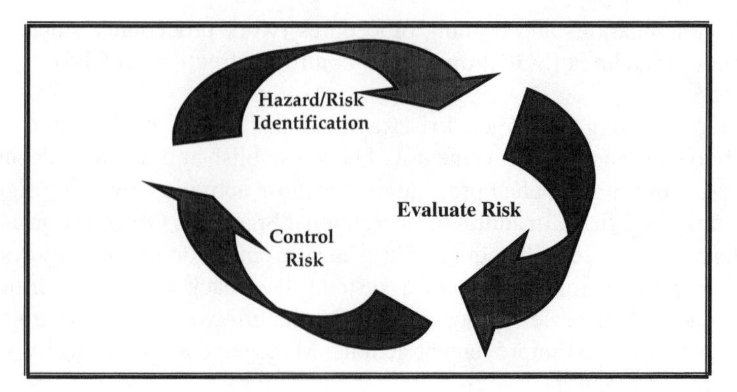

Figure 23 Hazard Identification, Risk Assessment & Risk Control.

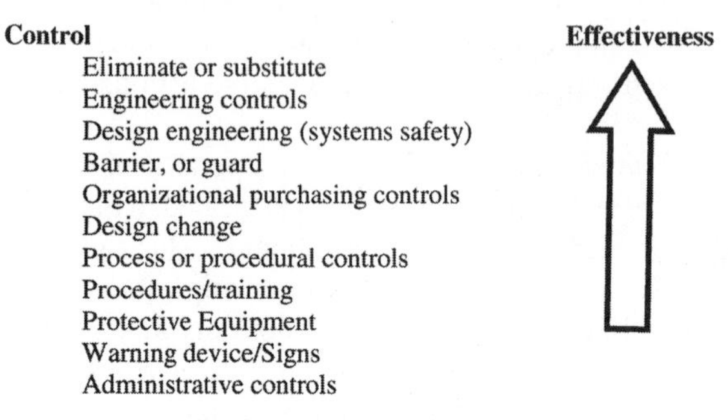

Figure 24 Hazard Control.

risks will require specialist advice, research, and information from regulators or industry associations to determine the best practices.

Inspection. A physical inspection of the work environment, tools, and equipment or an assessment of the work process.

Illness and Injury Records (Loss Source Analysis), Customer Complaint or Rework Records. Records of past incidents involving injury and illness highlight sources of potential harm or process nonconformance. Customer complaints or documented rework records flag process weaknesses and establish feedback to initiate appropriate corrective/preventive actions.

Risk Assessment, Critical Performance Observations, or Job–Task Analysis. By breaking a task down into its individual elements, risks associated with the task can be identified. Critical work processes should be evaluated from a readiness-risk perspective in relation to tools and equipment, facility, maintenance, training, supervision, and management and with the readiness/adequacy of these process resources inputs.

1. NAME			2. DEPARTMENT
3. OCCUPATION	4. JOB TASK OBSERVED	5. DATE	6. TYPE OF OBSERVATION INITIAL --- FOLLOW
7. TIME WITH COMPANY	8. TIME ON PRESENT JOB	9. NOTIFICATION PRIOR	--- NO NOTIFICATION
10. REASON OBSERVATION __JOB PROCEDURE/ PRACTICE UPDATE	--- PROCESS FAILURE --- ACCIDENT-INJURY	--- TRAINING FOLLOW-UP --- SIGNIFICANT TASK	--- KNOWN RISK --- QUALITY-EH&S RISK
JOB-TASK			
11. COULD ANY OF THE PRACTICES OR OBSERVED RESULT IN QUALITY OR EH&S RISK, PROPERTY DAMGE OR PERSONAL INJURY?	----- YES ----- NO	12. WERE THE METHODS AND PRACTICES OBSERVED THE MOST EFFICIENT AND PRODUCTIVE, AND THE METRODS REQUIRED IN PROCESS INSTRUCTIONS AND TRAINING?	----- YES ----- NO
13. DIDTHE PRACTICES OBSERVED COMPLY WITH ALL OF THE STANDARDS & TRAINING FOR THIS TASK OR JOB?	----- YES ----- NO	14. COULD ANY OF THE PRACTICES OBSERVED HAVE A NEGATIVE IMPACT ON THE QUALITY OR EH&S RISK ASSOCIATED WIHT JOB?	----- YES ----- NO
15. DESCRIBE ANY PRACTICES OR CONDITIONS RELATED TO ITEMS ABOVE THAT SHOULD BE REINFORCED OR			
16. WAS THE JOB-TASK PERFORMANCE OBSERVED REIMFORCED OR CORRECTED?	----- YES ----- NO	17. SHOULD A FOLLOW-UP OBSERVATION OF THIS EMPLOYEE OR TASK BE MADE?	----- YES ----- NO
18. DESCRIBE ANY PROCEDURES, TOOLS OR EQUIPMENT THAT WE SHOULD CONSIDER CHANGING TO IMPROVE QUALITY ER&S			
19. SUPERVISOR OBSERVER			20. DEPARTMENT
CORRECTIVE & PREVENTIVE ACTION FOLLOW-UP			

Figure 25 *Job–Task Observation.*

Hazard Analysis and Systems Safety Tools. Commonly used risk assessment tools such as Work Risk Analysis and Control (WRAC), Hazard and Operability (HAZ-OP) fault tree analysis, and Management Oversight and Risk Tree analysis (MORT) should be used to help assess the QEH&S risks associated with critical jobs/tasks and to determine prioritized cost-effective controls for at-risk procedures.

Audits. Include management system and regulatory or technical conformance/compliance audits or process control and follow-up audits. Compliance audits may be best accomplished by initiating cross-facility audits. This approach is not only an excellent way to complete the audits, but it also serves as a valuable tool for training managers and sharing information. The cross-facility audits will require both managerial and technical knowledge of the specific processes being audited.

To ensure a systematic evaluation of hazards, organizations should reference relevant quality and EH&S sources of information in the development and execution of these audit programs, including standards such as the following.

- Corporate policy, legislation, and EH&S approved codes of practice, which give practical guidance and include minimum requirements.
- Information provided by manufacturers and suppliers of tools, equipment, and materials.
- National and international standards.
- Industry or trade association guidance.
- Knowledge and experience of managers and employees.
- Accident, illness, and incident data from within the organization, other facilities, or similar industries, including customer complaint or rework records.
- Expert advice and opinion from competent process QEH&S staff or consultants.

Processes and equipment must have documented compliance control, including administrative procedures defining the production, installation, and servicing of the system. Control procedures are necessary to prevent a breakdown of critical process QEH&S. In addition, a formal review and approval processes for procurement or the designation of materials, tools, and equipment for critical jobs or process tasks must be in place.

Process design and SOPs should consider maintenance of equipment to preserve continuing QEH&S system performance. Maintenance review schedules should also assess controls such as guards, noise barriers, and ventilation systems. The management system should establish procurement procedures to facilitate potential material substitution of hazardous materials with less hazardous materials or modification of the process design to eliminate risk.

PURCHASING

Purchasing and Engineering Controls. Process redesign will be less effective, and much more expensive than good job or process planning (systems safety). Designing controls before equipment or materials are bought on-site is usually much less costly. Systems safety engineering should be used when designing new facilities or processes. QEH&S risks associated with vendor-supplied goods must also be considered. Purchased products must conform to customer or process specifications, and the EH&S risks associated with the shipping, storage, and use of hazardous materials should be communicated to management and employees. Procedures should include obtaining and distributing material safety data sheets and reviewing incoming material to verify receipt and conformance to order specifications of hazardous material, such as inventory control, storage, shipping, and spill containment requirements. Organizations must establish and maintain documented procedures for the review of QEH&S specifications for goods and services, including ensuring that vendor materials meet applicable requirements before being accepted for production or process use.

In many countries this process is dictated by both environmental and safety laws. The system must have cradle-to-grave material use and inventory tracking capabilities. This process requires evaluating and documenting procedures and the chemical inventory from the HM procurement purchasing review and receipt stage to ultimate use and waste/disposal or recycling stage. Purchasing documentation should be maintained to accomplish the following.

- List preferred and approved suppliers and contractors.
- Show the decision-making process, including risk assessment and product specification or verifications for receipt of purchased goods and products.

A system for the purchasing of goods and services, including maintenance procedures, is a key element of any organization's process enhancement strategy. The system should ensure that purchased goods, services, products, and contractors conform to the organization's QEH&S requirements. Personnel performing these reviews must have documented skill and QEH&S specifications.

Documented procedures should be established to ensure that technical specifications and other relevant QEH&S requirements are examined prior to purchasing decisions. In addition, the qualifications of vendors and contractors to deliver quality products and services must be evaluated and documented, including approved vendor/contractor lists. This process should include these items.

1. A review of contracts and orders to ensure customer requirements are adequately defined. Any differences in material availability or received materials are mutually resolved with the customer or operations manager before using the substitute material.
2. The capacity of the vendor to fill the order on time and within budget should be determined before placing orders or signing service contracts with them.
3. Procedures for contract amendments or changed orders must be defined and documented.
4. Subcontractors' and vendors' credentials need to be verified prior to entering a contractual obligation with them. Procedures for on-going assessment of work during the contract period is also necessary to ensure QEH&S conformance with specifications. It is recommended that facilities maintain a list of approved contractors, including documentation supporting their QEH&S performance.
5. When appropriate, it is worth while to verify purchased products at the sub-contractor's facility and to offer the same courtesy to your customers.

Product Identification and Traceability

Identify critical materials, products, procedures, equipment, or resources to be used during all stages of production or service delivery so that changes in any of these process components are not made without approval of the project or work center manager responsible for its outcome.

Identify, tag, or label all unique products, batches, or tools required for product outcome to ensure its traceability and protect customer-supplied tools, procedures, or products that will be used or incorporated in the delivery of the service or product. Customers must also be notified if these resources are lost or damaged.

Establish a systematic process to identify and segregate nonconforming products, subassemblies, or components of a service or service delivery report, and segregate it until it has been changed to conform or is replaced with a conforming product or service plan. After determining the multiple root causes of all product or service nonconformances, implement appropriate corrective/prevention actions, including changes to procedures and training.

EH&S risks related to purchased goods and products should be communicated to management and employees, including obtaining, reviewing, and distributing material safety data sheets, and the inspection of incoming material to verify receipt of materials in accordance with procurement and/or customer requirements. Hazardous material inventory control, storage, shipping, and spill containment requirements must be considered and established where there is a risk of releasing hazardous materials.

A system for purchasing goods and services, including maintenance procedures, should be integral to an organization's process enhancement strategy. The system should ensure that purchased goods, services, and products and contractors/subcontractors conform to the organization's QEH&S requirements. Personnel performing these processes must have documented skill and experience and must follow procedures confirming that purchased goods and services conform with QEH&S specifications.

Facilities should establish and maintain documented procedures for the review of QEH&S specifications of goods and products. Goods and products must meet applicable QEH&S requirements before being accepted for production or process. Procedures must include requirements to inspect or verify product receipt and to prevent its use if the material does not meet design, contract, purchase specifications, hazardous material specification, or product use requirements.

Contract Review

The following approval process should be documented with every vendor or contractor, including documentation of the contract review.

- QEH&S requirements are adequately defined and documented.
- Any QEH&S requirements differing from those in contract or customer specifications are resolved.
- The organization (i.e., the supplier) has the capacity to meet the QEH&S contractual requirements.
- Records of contract reviews should be maintained.

- Contracts are reviewed to ensure that the organization is capable of meeting the identified QEH&S requirements of its customers and that potential risks are identified and evaluated prior to starting the contract.

Management of Contractors and Suppliers

The organization must have a policy for the employment of contractors (and subcontractors). This document should establish a procedure to ensure that its policy, plans, and procedures for contractors have been communicated to line managers, supervisors, and employees. Some organizations may choose to maintain a register of preferred contractors that have established and maintained effective QEH&S systems and practices. The organization should select contractors based on their ability to meet the organization's QEH&S requirements. A contractor's ability to meet these requirements can be assessed in accordance with these factors.

- The organization's QEH&S policy.
- Project work plans.
- Registration and licenses (where applicable).
- Agreement to comply with the employer's QEH&S requirements.
- Verification (by inspection and tests) that work areas, work methods, materials, plant, and equipment comply with all QEH&S requirements, legislation, regulations, standards, codes, and product/process quality specifications.
- Identification and allocation of resources to meet specified requirements.
- Pre-planning — for medium/long-term contracts, this step could involve full QEH&S prequalification/evaluation procedures. For short-term contracts QEH&S could be checked by questionnaire or review.
- Defined responsibility and accountability for establishing communication links between organizational levels and the contractor prior to contract commencement.
- QEH&S training of the contractor personnel where necessary before commencement of work.
- Monitoring QEH&S aspects of the contractor site activities.
- Communication and reporting procedures for accidents and incidents involving the contractor's personnel.

Contractors should be capable of demonstrating their commitment to QEH&S and how they will comply with the organization's requirements. In addition, subcontractors and suppliers should be selected on the basis of their ability to comply with QEH&S requirements, including process verification procedures such as inspection, testing, and auditing.

Design and Engineering Control

Facilities must identify and document their plan or procedure for process design reviews and ensure that competent and experienced personnel use systems safety tools, such as MORT or FMEA, are appropriate for critical or high-risk tasks. All design reviews must be documented and changes or corrective actions tracked to completion before executing the final design change and approval process. In addition, there should be a formal procedure for reviewing and giving signature approval to design or process changes. The system should formally document customer or design output requirements and ensure that system changes verify that customer requirements have been met.

Procedures and work instructions relating to the qualified and the safe use of products and processes should be developed during the design stage of new processes. In a paper presented at the annual ASSE conference in Baltimore in 1999, Sandra Bissett from Gillette discussed how the company estimated that it increased the design, process, and equipment installation by 20% but prevented even greater costs of inadequate planning, such as start-up delays and downtime.

Product or Service Design and EH&S Considerations

- Customer requirements
- Construction and maintenance of plant and equipment
- Design and layout of work areas
- Equipment design
- Ergonomic principles
- Product liability
- Ensuring legal standards are met
- Job design and work procedures/systems of work
- General purchasing
- Environmental management
- Purchasing, use, storage, transportation, and disposal of chemicals
- Arrangements with suppliers, contractors and subcontractors

Companies must plan production processes and provide personnel with procedural control instructions, equipment, material, and tools that will ensure product quality and the safe handling, storage, and packaging of HM/HW. Storage includes storage of raw materials, in-process product and finished product, and the temporary storage of HW (shipment, disposal or treatment).

Organizations should require that QEH&S decisions relating to modifications to buildings, processes, tools, equipment, materials, or service changes are approved by competent personnel. Records of design verifications, modifications, and approvals should be maintained. Systems safety, risk assessments, and failure mode analysis should be used to evaluate critical processes or planned operations.

Each stage of the design cycle (development, review verification, validation, and change) should include hazard identification, risk assessment, and risk control procedures. Procurement and design efforts must also consider alternatives for HM/HW, for example, the use of less hazardous material, spill contingency, and waste minimization or recycling. Where new process risks cannot be eliminated or substituted for one that presents a lower risk, engineering controls should be included in the product production process design.

Design, Siting, and Selection of Premises. Prospective process risks may be identified at the design stage in consultation with competent professionals, engineers, architects, doctors, or health and safety professionals. The risk associated with these hazards may be controlled by QEH&S risk control considerations at the initial design and planning phase of the process. The intended use, as well as maintenance of facilities, equipment tools and systems should be considered when conducting risk assessments for these key processes.

If a process, product, or workplace is designed and built with QEH&S in mind, the number of reactive add-on controls or procedures (controls that are costly and often times less effective and more difficult to maintain) will be minimized. The proposed use, possible future uses, and maintenance requirements should be consideration factors in site or facility design and selection. In addition, property owners should be concerned about the potential environmental or HW issues associated with the site or adjacent site's current and historical activities.

Managers should establish and maintain documented procedures to control and verify the design of facilities, equipment, processes, and systems to satisfy company policy and any relevant regulatory requirements. Design or procedural control instruction changes that modify or potentially impact product QEH&S must be reviewed, including compliance with applicable standards, and approved by competent QEH&S personnel prior to implementation.

Organizations should require that competent personnel approve QEH&S issues at the design stage. Design should incorporate as appropriate hazard identification, risk assessment, and risk control procedures. QEH&S must be represented in production or production process engineering planning and design and process or procedural development efforts, including ergonomic considerations.

PROCESS IMPROVEMENT—RISK REDUCTION

Part of continual improvement will be to develop a system to validate and approve new process designs or procedural changes. Organizations should establish and maintain documented work procedures that contain QEH&S best practice procedures for activities with significant QEH&S risk. This standard, in turn, will require employees and supervisors to be trained to use these procedures.

Effective QEH&S management will require that facilities establish and maintain procedures to ensure that new or recently modified equipment, tools, facilities, or processes have design/operational characteristics that are consistent with their intended use and meet necessary QEH&S performance specifications.

Maintenance procedures will also be necessary to ensure the safe operation of plant equipment, processes, and services and to ensure that employees always have tools and equipment designed for or intended to be used for specific work processes. These process and document design reviews should be performed by competent QEH&S personnel.

The approach for controlling QEH&S risks should adapt the following preferred hierarchy of controls from elimination (most desirable) to use of personal protective equipment (least desirable).

Elimination/substitution: This solution is permanent and should be attempted first. If the hazard is either eliminated or substituted by one that presents a lower risk, additional design or control changes may not be necessary. This objective could involve the elimination of a hazardous process or substance or the substitution of hazardous materials with a less hazardous material.

Engineering controls: These controls involve some structural change to the work environment or work process such as placing a barrier between the worker and the hazard. It may also include machine guards, isolation or enclosure of hazards, and the use of exhaust ventilation and material handling devices.

Personal protective equipment: This equipment provides a barrier between employees and the hazard. The success of this control depends on the effectiveness of how the protective equipment is actually worn; its selection, maintenance, and fitting; and the medical or physical ability to use protective equipment.

Administrative controls: These controls reduce or eliminate exposure to a hazard by adherence to procedures or instructions. Such controls may include reducing the exposure to a hazard by job rotation, permit-to-work systems, or they may simply involve the use of warning signs. Administrative controls depend on appropriate human behavior for success and therefore should generally be the least desirable control options.

Attempts should be made to develop control measures from the top end of the hierarchy wherever possible. However, it may be necessary to use a combination of control measures to achieve the desired level of risk control (see Fig. 24).

Organizational or Facility Systems of Work Guidelines

Procedures should be documented and, where necessary, permit-to-work systems for high-risk tasks should be in place. Risk control measures should be reviewed when there is a change to the work process. In addition, compliance with relevant regulations, standards, and codes of practice should be periodically reviewed when developing or modifying procedures or work instructions. Work procedures and work instructions should be developed by competent persons with input required from employees performing the task, customers, or other stakeholders as appropriate.

Supervision Guidelines

Supervisors are the heart of any efficient organization, and they ensure that tasks are performed safely and that process instructions are used. Supervisors should also participate in hazard identification, risk assessments, and the development of associated control measures. They should be trained and should participate in injury, illness, and nonconformance reporting and investigation, including preparing required reports and recommendations for senior management (see Chapter 8). Supervisors should also actively participate in their work center's PETs.

Maintenance, Repair, and Alteration of Plant Guidelines

A maintenance request procedure should include provisions for unsafe tools so that they are withdrawn from service (lock out/tag out may also be required). Inspection and maintenance of plant and equipment should be scheduled and should include verification of control devices and the implementation of QEH&S requirements, legislation, standards, and codes of practice. Records should be maintained documenting the details of inspections, maintenance, repair and alteration of tools, or equipment. A sign off or certification procedure should also be in place to ensure that tools or equipment being returned to service conform to design/operations requirements.

Materials (Use, Storage Considerations)

To reduce costs and lower the risk of a chemical hazards, management should reduce its excess chemical inventory and, wherever possible, substitute for less hazardous materials. As part of the community and employee right-to-know programs, requirements should be developed to ensure that all-hazardous materials containers, tanks, and pipes are labeled and that employees are trained to understand the risks, required controls, and emergency response procedures for working with or around hazardous materials. Hazardous materials business plans (community right to know) should be filed with the fire department or designated community emergency response agency, and, where appropriate, these outside agencies should participate in facility emergency response drills. To the extent possible just-on-time delivery and material inventory management practices should be used to reduce staging costs and other risks such as shelf expiration.

Hazardous Material/Waste Management

Organizations should establish and maintain documented procedures for the procurement, handling, storage, packaging, transportation, or treatment of hazardous or potentially hazardous materials and waste. Storage includes storage of raw materials, in-process product, finished product, and the temporary storage of hazardous waste (shipment, disposal, or treatment). To minimize the risk of handling, storage, and disposal of HM/HW, facilities must consider alternatives for these materials, including substitution, material treatment, and recycling or reuse of hazardous materials and waste.

Storage facilities should establish procedures to handle, transport, ship, or store materials or products to prevent damage or the accidental release of hazardous material. Procedures could include product-packaging requirements, containment, material segregation, and spill response. Procedures should also include routine inspection of products in storage or materials to be used in the production process for quality and usability. As appropriate, some hazardous materials, waste, or damaged products must be segregated. Inspections of storage areas should include fire protection and the compatibility of stored materials, chemical storage containment to prevent accidental release of hazardous materials, and regulatory permit-reporting and emergency response requirements.

HM/HW Transport Storage and Disposal Procedure Recommendations

- Procedures must ensure materials are stored and transported in a safe manner in accordance with legal requirements and documented procedures, including shipping manifests, packaging, labeling, and the use of transport placards in accordance with Department of Transportation (DOT) requirements.
- Procedures that specify requirements for the control of materials subject to deterioration or a "use by" date.
- Requirements to ensure materials are disposed of safely and in accordance with regulation.
- Comprehensive material safety data sheets (MSDS) for HM must be readily available.
- Systems for the identification and labeling of hazardous materials.
- Training employees for handling hazardous materials to ensure safe practices.

Employee Hazard Communication. Employers are responsible for evaluating chemical hazards in the workplace and must supply information to employees in the form of a written hazard communication program. Facilities must also ensure that inventories of chemicals used and their MSDS are available for all potentially exposed employees. Likewise distributors of chemicals must relay information to their customers regarding the hazards, use, and controls specified on MSDS.

Process, equipment, and tools should be periodically inspected to ensure conformance with QEH&S requirements and that maintenance is performed as specified in facility or manufacturing procedures. Tools or equipment requiring maintenance or replacement should be removed in accordance with documented procedures until the required maintenance and/or calibration is performed.

TRAINING

> Most of the early problems with this unit (to show sound film) were due to an inadequate instruction book, translated from the German into English by someone illiterate in both languages.
>
> —*Bulletin of the Washington Society of Cinematographers, November 1967*

The effective implementation and maintenance of a QEH&S management system is dependent on staff competencies, training, defined procedures, and effective aligned communications from supervisors and management. Training is an important means of ensuring the appropriate competencies to achieve QEH&S

objectives. Procedures should be in place to define competency standards and to evaluate individual training and coaching needs. Personnel should be trained to understand the QEH&S policies of the facility, including compliance roles for achieving defined objectives and milestones.

Personnel assigned to perform tasks that could affect the quality of products or services or the EH&S associated with their assigned tasks should be competent, in terms of education, training, or experience, to perform their duties. All personnel (management, staff, or contractors) who can potentially impact QEH&S performance as it relates to at-risk processes or significant identified procedures must be trained. The training objective is to understand the risk associated with individual activities and the procedures that must be followed to mitigate risk or control the work process. Training should also include emergency preparedness and response procedures, basic process control, quality, and QEH&S systems management.

Supervisors and managers must also be trained to perform process failure/nonconformance and accident investigations with an emphasis on understanding multiple root cause analysis and sharing lessons learned as a key to corrective and preventive action implementation. The integration of QEH&S into process instructions, regulatory conformance, emergency procedures, and contingency plans should also be included in the training for management and employees at all levels.

> [The Japanese] took an American invention of World War II—it was called *training* and it enabled the United States during the war years to change preindustrial, unskilled people into efficient, high productivity workers—and turned their unskilled and low-wage people rapidly into highly productive but still low-wage workers whose output could then compete in the developed markets.
>
> —*Peter F. Drucker,* Managing for the Future, The 1990 and Beyond

Procedural Controls

The first step to enhanced safety and quality is to develop written job procedures or practices. Procedures are teaching tools and they can be used to measure performance during inspections, audits, or incident/accident investigations.

Quality and EH&S competency standards can be developed by the following methods.

- Examining job or position descriptions
- Analyzing work tasks (job–task analysis or risk assessment)
- Analyzing results of inspection and audits
- Reviewing incident reports (customer complaints, rework, or accident/incident reports)

A training program, including individual development plans, should be developed after an assessment of current capability and the required competency profile of individuals and the organization are completed. QEH&S training should be integrated as part of individual and organizational development plans/budgets. Procedures should be established to document training and to evaluate its effectiveness.

A training program may need to address a number of target groups, including the following.

- Senior management
- Line managers/supervisors
- Employees (new or transferred from other jobs)
- Personnel with specific QEH&S, quality measurement or benchmarking, and emergency response responsibilities (first aid, fire, health and safety)
- Contractors
- Operators, managers, and technical staff who require certificates or certification maintenance
- Site visitors

Training records should normally include the goal or purpose of the training, including how it satisfied corporate and individual development and regulatory compliance information about the trainees and course should include the following.

- Material covered in the training course
- When and where the training took place
- Who provided the training
- How often refresher training or recertification should be required

Training Strategy

Training should be provided by experienced trainers, experts, or professionals in QEH&S versus the common practice of train the trainer. Training should be experienced-based by using stories and metaphors, group discussion, and workshops to the extent possible. Demonstrations with real case studies to illustrate processes are always excellent ways to reinforce learning and to facilitate group discussions so participants can share their experience and lessons learned. A training-needs analysis should be conducted, which would includes QEH&S requirements, and individual development plans (IDPs) should be documented for each employee. Other training program considerations should include the following.

- Training programs must consider different levels of ability and literacy.
- Facilities and resources must be suitable to deliver effective training.
- Every training session should be evaluated to ensure comprehension and retention, and to ascertain that instructors met the goals and expectations of the attendees.
- Training programs should be reviewed regularly to ensure their relevance and effectiveness.
- When changes to processes in the workplace take place, training should be provided to affected personnel.

- Refresher training must be planned for all personnel as appropriate, including documented job/task observations to confirm the competent execution of critical work procedures.

Quality and Environmental Health and Safety Procedures

Where other means of control have been ruled out, documented procedures should be established and integrated with existing quality or environmental management systems.

Procedures and work instructions should be written with safe, efficient productivity integrated into each process step. The design and review of work procedures should be developed by competent people and those involved with the product or production design or management of the tasks.

Procedures or procedural control instructions should be reviewed regularly, as well as when changes to equipment, tools, processes or raw material have occurred. Management should establish maintenance and review schedules for critical QEH&S procedures to ensure that they are current with regards to applicable regulations and industry best practice standards or customer requirements.

Administrative Control Procedures for Consideration

- Rotation of jobs, for example, to reduce exposure to a specific hazard such as noise.
- Permit for high-risk tasks, for example, isolation/entry, confined space entry, lock out/tag out, or hot work. Permits must be signed and dated.
- Restricted work areas, for example, for welding or use of explosive-powered tools.
- Use of warning signs, labeling.
- Implementing a maintenance program for plant and equipment.

Emergency Preparedness and Response: Critical Incident Recovery Plan

The organization should determine acceptable levels of risk by reference to legislation governing health and safety, standards, codes of practice, or industry best practice. Risk assessments should consider the likelihood of an undesirable or unplanned event (nonconformance/accident) and the potential severity or magnitude of the process failure.

Where the measures to control process risk are not considered adequate, the evaluation can be used to set priorities for any remedial action. Organizations should require that QEH&S issues be considered in decisions relating to modifications to buildings, processes, tools, equipment, use of materials, and procurement procedures. Process changes must be reviewed and approved by competent personnel. Process review documentation should include design verifications, modifications, and approvals. To assist with the review process systems safety, risk assessments, or failure mode analysis can be used to evaluate and document critical processes or planned operations.

Each stage of the design cycle (development, review verification, validation, and change) should incorporate quality or process risk and hazard identification, risk assessment, and risk control procedures. Procurement and design efforts must also consider alternatives for hazardous materials and waste, for example, use of less hazardous material, spill contingency, and waste minimization or recycling.

Emergency Preparedness and Response. The organization should develop documented emergency response plans and procedures, which establish the procedures in the event of an emergency. Emergency plan procedures that must be periodically tested and should be reviewed by service providers, such as the fire department or the police. For large installations, the emergency plans may need to be coordinated with municipal or state disaster planning. Emergency preparedness and response plans should address the potential for natural disasters and facility or transportation risks. All identified emergency conditions should have well-defined response plans, including drills and community emergency response support procedures as appropriate. Emergency organization responsibilities should be defined and understood by management and those responding to emergency situations. Facilities must also install warning and alarm systems and ensure that they are periodically tested.

Emergency instructions and emergency contacts must be prominently displayed and known throughout the organization. An internal and external communication plan should be a key component of the emergency response plan and should include a list of key personnel, including emergency contacts/recall lists; procedures for involving community support organizations; and public relations, as necessary.

Employees involved with emergency response (e.g., fire department, spill clean-up services, etc.) must receive specialized emergency response training and be provided with the necessary equipment to readily respond to emergency conditions. In addition, these tools, procedures, and training should be periodically inspected and maintained.

All drills or emergency response actions should be critiqued by senior managers and emergency response personnel to identify the need for process improvement due to observed nonconformance or inadequate/partial response to emergency procedures. Emergency procedures should be regularly tested and reviewed by competent personnel, including management and community stakeholders.

Hazardous Material/Waste Spill Response Plans

These response plans must contain detailed inventories, including the following.

- MSDS with chemical and common nomenclature of every hazardous material or waste handled.

- Category of HM/HW, including chemical and mineral composition.
- The maximum amount of each material on site at any one time during the year.
- The storage location and information for fire, safety, health, and other appropriate agencies to prepare for emergency response, including site maps identifying the location of storage areas and the volume of hazardous materials stored there.

Medical Surveillance Programs

An integral part of a QEH&S evaluation system should include a medical program. Having a medical program does not necessarily require having an on-site nurse or doctor. It does mean involving occupational health professionals in work-site analysis of hazards and in hazard prevention and control programs. It also involves early recognition and treatment of illness and injuries, post-injury return-to-work programs, and the limitation and severity of repeated injury associated with workplace accidents.

Some organizations elect to develop and implement medical surveillance programs such as pre-placement examinations, return-to-work examinations, and substance-abuse testing. Where these programs have been instituted, they should be developed and supervised by qualified individuals such as occupational health nurses or occupational health physicians. These factors should also be considered when evaluating the adequacy of work-site medical programs.

- Processes and conditions under which employees work
- Types of materials employees used
- Types of facilities and distance to health care facilities
- Number of employees at the work site
- Characteristics of the workforce such as age, gender, education level

Action Plan

- Develop policy and process for facility or process design control and review, including PCI or SOP development.
- Establish management, supervisor development programs, and employee QEH&S and hazard communication training.
- Establish emergency response plan.
- Develop risk assessment–contingency planning process.
- Evaluate and risk rank critical tool and equipment inventory/maintenance program.
- Establish work center PETs and Qualitative Risk Assessment (QRA) procedures.
- Establish procurement/contracting review procedures.

- Establish maintenance schedules for critical tools, equipment, and operations.
- Institute IDPs.
- Establish HAZCOM/chemical management system, including storage, requirements, review of material compatibility, permit, requirements, secondary containment, and labeling.
- Develop SOP for critical job–tasks.

7

Measurement and Evaluation

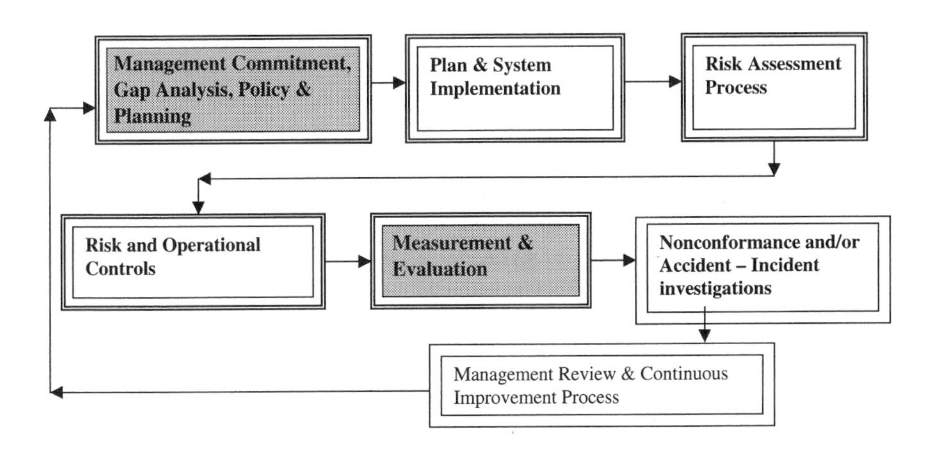

It is more than probable that the average man could, with no injury to his health, increase his efficiency fifty percent.

— *Walter Scott*

Any man can make mistakes, but only an idiot persists in his error.

— *Cicero*

What gets measured gets done. Measurements are like GPS readings; a sailor returning home in the fog can use it to stay on course and return home safely or risk seeing the organization on the rocks. In this chapter we will discuss the use of several measurement or evaluation tools, including audits, inspections, job–task analysis, performance management/behavioral safety, and a brief discussion of statistical process control (SPC). (Note that SPC is discussed in more detail in Chapter 5.)

Managers need proactive measurements of performance by using audit and critical performance observation tools to identify system nonconformances and process risk before process failure occurs. This chapter discusses these tools and how to use the proactively generated information for periodic management review and the continuous improvement process. It also discusses life cycle risk analysis, audits, and inspections by using defined audit criteria, inspection, or critical job observations with check sheets.

Managers must set measurable goals as well as the policies/procedures to execute those goals. Once you sail out of the harbor you will need to periodically check your course and make the adjustments to the system to optimize the most cost-effective implementation of an integrated management system. Quality and Environmental Health and Safety (QEH&S) managers must also be able to communicate with managers in financial terms, considering factors such as these.

- Payback period and return on investment
- Time value of investments or proposed corrective actions (present value–future value calculations, discounted cash flow, etc.)
- Ability to document the direct and indirect cost of process
- Nonconformance

Operations management must be in a position to financially evaluate risk and the cost/benefits of proposed control alternatives as compared with requests for other limited corporate resources. Typically, management will need to know what the total annualized costs of nonconformance are and the expected benefits of the proposed corrective alternatives. In addition, management may need to consider post-product service liability as the product waste is moved into waste treatment facilities or landfills.

In Chapter 5 we discussed the concept of life cycle risks and provided a chart that briefly illustrates the potential generic risks confronting many facilities in the product manufacturing cycle. The integration of QEH&S into management decision making will require these managers to consider the following issues to effectively enhance performance, communicate their needs, and compete for limited resources.

1. How can the company better measure and report QEH&S risk and improve decision making?
2. How can an integrated management system (IMS) be institutionalized and measured by the same criteria as other management systems or departments?
3. How can QEH&S be better integrated/recognized as a component of product–service pricing, capital investments and operational performance?

4. Data will be needed to quantify the cost reduction in rework and waste (including hazardous waste).
5. Existing QEH&S liability costs will be needed, including documentation by both product and facility (past product rework or process nonconformance, environmental releases, and employee injuries, waste liability, etc.).
6. A process to document direct and indirect costs of process nonconformance will be needed, including charge back of these costs to the nonconforming work centers.
7. An estimate of both short-term and future capital expenditures for QEH&S control, and nonconformance prevention and/or product redesign to reduce costs and product life cycle risks.
8. Estimate of future costs–benefits of QEH&S and product life cycle risk management.

According to the Business Charter for Sustainable Development (BCSD) of the International Chamber of Commerce, management accounting systems must include the environmental costs that are part of the production, distribution, marketing, use, and disposal of goods.... The BCSD seeks a full cost accounting that takes into account internal and external costs. (Epstein, 1998)

Information system (IS) support will be needed to cost-effectively and accurately document nonconformance incidents and their total cost, including the cost–benefits of control alternatives. These data should also be used in product pricing and product-process design considerations. IS will also provide data for management system and regulatory compliance. Continuous improvement will be difficult without tools to document cost and process information, including the lessons learned with other work centers and departments in the company.

With this kind of documented cost–benefit analysis, improved product pricing and QEH&S decisions can be made relative to capital investments and product design changes, which will also improve capital budgeting and performance evaluations. QEH&S considerations must be a critical component of strategic and financial management decisions. The risk considerations and control alternatives, their implementation, their success measurements and follow-up must be critical elements of performance evaluations, product design cost, and capital investment decisions.

Corporations have been underestimating and overestimating the costs of procedures and their alternatives based on inadequate information regarding costs and benefits of corporate environmental (EH&S) impacts. Improper costing and capital decision making have cost industry billions of dollars. (Ref. 20)

I will also encourage you to consider understanding the value of prospective measurements by using performance management tools, otherwise you will be forced to drive forward while looking in the rear view mirror. This analogy is often especially true if you rely on financial data, which is 6–8 weeks old so that it is difficult to prevent errors or substandard performance until a significant period has passed. Retrospective financial and operational performance

TABLE 13. Examples of Life Cycle Cost Categories

Conventional Costs	*Liability Costs*	*Environmental Costs*
Capital	Legal	Global warming
Equipment	Penalties and fines	Ozone depletion
Labor	Personal injury	Photo chemical smog
Energy	Consumer-employee tort actions	Acid rain
Monitoring	Product liability	Water pollution
Regulatory compliance	Workers' compensation and other injury–illness costs	Resource depletion
Documentation	Remediation activities	Chronic health effects
Maintenance	Loss of market share, adverse PR	Habitat alternation
Air emission control	Share holder loss in stock value	Social welfare effects
Water effluent control	Cash flow impact due to establishment of required reserves/contingencies	
Radioactive/hazardous material control	Property damage	
Waste management	Business interruption	
Raw materials and supplies		
Waste treatment/disposal Costs		
Spill containment costs		
Emergency response Planning and drill costs, including community interface		

Source: Ref. 20

measurements are needed; however, as you gain control of your management system you need to be focusing on the road ahead of you, and not the past.

What happened yesterday is difficult or impossible to correct. The distinction that must be made is that traditional financial and operational management tools focus on results or the past. Quality management systems focus on real time and on the process the road ahead. You are therefore able to better control process risk, thus ensuring the delivery of quality products, without exposing stakeholders to EH&S risks. Effective QEH&S also requires relevant and forward-looking objectives to measure system or process performance against.

Management by objectives (MBO), production rates, and quotas are usually set by taking into consideration the average employee. The fallacy of these approaches is that they fail to recognize that half of the employees will fall below the average and the other half, above it. This tendency results in a cap on production, fear from employees who are below average, and in some cases pressure on the employees who can exceed average to not go beyond minimum requirements.

MBO, as practiced by most organizations, fails to recognize the interdependencies between departments within an organization and as a result tends to foster unhealthy competition versus cooperation. Unlike an approach to improve the system, a system that rewards those who do well focuses too much on short-term individual results. This tendency in turn stifles team efforts to improve the system or processes, self-initiative, or risk-taking. A better individual and organizational performance evaluation system is the one used by Lincoln Electric, which is discussed in several places in this book, including Appendix III.

Cooperation and teamwork comprise a major point in the discussion of fear in Deming's Eighth Point, which emphasizes a need to break down barriers between departments and staff areas. One factor that works against teamwork is the tendency for departments or functional units to strive for optimization of their own work and to disregard, either intentionally or unintentionally, relationships that their work has on others. This problem unfortunately tends to get worse the farther removed staff functions are from the "point of the lance", as some of my European friends put it. Or when the employee is distanced from the external paying customer (for whom the entire organization is ultimately in place).

The objective is to focus all processes to provide customers, whether they are external or internal, with the delivery of a product or service they expect. The organization as a whole must coordinate its internal procedures and activities to support the sales, marketing personnel, and production managers who are directly involved with providing the external (paying) customer with the services and products they have requested and paid for. In addition, activities that don't directly support product or service development and delivery or QEH&S should be evaluated to determine if it is needed or if it supports the goal of quality delivery of goods and services to the paying customer. The more redundancy and process QEH&S risks we can eliminate from the system, the easier it will be to profitably deliver to the paying customer. I have too often seen large organizations where support managers and staff forget or don't understand who are the "Point People" or operations managers and staff.

This can be a critical management issue. Support staff and managers only exist to support operations, or the people who directly interface with and support the "paying customer". Without this understanding there will be a lack of focus, and administrative errors that compromise the operations managers ability to consistently support the "paying customer".

I am sometimes amazed over how this is misunderstood. A simple test is to identify your customer, and if they do not pay a check to the organization you are a support person or manager existing to support operations. The support team should always be asking themselves what they can do to better support operations, improve reporting, efficiency and quality.

The bottom line is that the operations managers work for the paying customer. They earn the revenues that pay the salaries and expenses of the support managers and staff who will not be needed if operations fails due to lack of support. Support staff are simply further down on the "food chain", and work to support the operations team. Unfortunately the desire for power and ego driven motivation too often cloud this reality.

Everyone in the organization must understand who the paying customer is, what he or she expects, and what we provide that will be most useful and economically delivered. Quality is customer satisfaction, and the key to prospective management is to develop processes and critical task observations that support the customer delivery process. The process must be efficiently executed by eliminating waste and focusing on continually improving the process of meeting or exceeding customer expectations, while not exposing workers or other stakeholders to potential EH&S hazards.

GOALS OF MEASUREMENT AND EVALUATION SYSTEM

Evaluation systems allow you to identify critical tasks and evaluate the effectiveness of the process design and systems support to ensure that substandard conditions are not cause for substandard employee performance or behaviors. Measurements and evaluations tools should include the following.

- Evaluate and help improve existing standard operating procedures (SOPs)
- Measure performance against those standards
- Evaluate performance and measure results
- Identify where you must correct substandard performance
- Help managers identify and commend/reinforce desired performance
- Evaluate the management tools for the system review and continuous improvement process
- Results of process or operation risk assessments
- Investigation reports for accidents/incidents to determine the multiple root causes of the process failure
- Environmental accidents, HM/HW releases, rework, or any other recognized nonconformance with process or regulatory requirements
- Results of operational risks assessments, including security, transportation, or political risks
- Customer complaints and/or post sales or delivery service issues, product, or service rework reports

Companies must consider and evaluate the measurements of data from all areas identified above to have an effective process enhancement or risk. Management must also avoid being punitive, or the data collection process will be flawed generating an atmosphere of fear and intimidation. The bottom line is that management must continually evaluate tasks and procedures to ensure a continual improvement process.

Review and Evaluation

Executive management reviews of the management system should be performed at defined intervals to ensure its continuing suitability and effectiveness in satisfying the organization's QEH&S objectives. Assessment outcomes should then be incorporated into periodic management reviews and corrective/preventive action planning processes. This process should also include assessments to enhance efficiency and productivity.

To the extent possible quality and EH&S management systems should be integrated, and the organization should consider organizing and maintaining a documentation summary for system required documentation. The QEH&S policy, objectives, and targets should have defined, reasonably achievable, and date-bound plans describing the means of accomplishing QEH&S objectives. Documented plans should define key roles, responsibilities, and procedures or provide description of other actions required by the organization's management system. Where appropriate, documentation should include the verification of action item implementation of applicable management system requirements. Other documentation should include records such as the minutes of QEH&S management review meetings.

Planned Inspections and Audits. Substandard conditions, in addition to being a direct cause of process failures, can lead to makeshift equipment or process changes and other substandard process performance. These conditions can develop in any area of operations. They result from factors such as these.

- Normal wear and tear
- Misuse and abuse of equipment and tools, including inadequate maintenance
- Poor communications, including supervision, procedures, and training

Planned inspections provide a systematic examination of work practices, facilities, equipment, tools, and materials used. They also analyze substandard practices or conditions contributing to risk potential. When the multiple root causes of process failures have been determined, the process can be changed to prevent future incidents or failures. To be successful, organizations should establish an experience and/or training and re-certification process for auditors.

Planned Inspections

A. They identify and classify sources of risk.

B. First-line supervisors and senior management teams inspect job sites for compliance with procedures or to evaluate the adequacy of procedures.

C. Inspections include the following.

Checklists of conditions and worker practices

Systematic evaluations of work flow

Identification of critical parts/materials/equipment (including maintenance) that could injure personnel, contaminate the environment, or result in production delays and rework etc.

D. They identify and classify risk.
E. They cause supervisors to initiate corrective/preventive actions for *potential* identified risks. Their inspection report must include a target date for risk abatement, documentation of interim controls, and ultimate risk prevention methods to be executed.
F. They identify serious process risks, helping the supervisor to initiate plans for interim and permanent corrective/preventive actions.

Critical Parts Inventory. A critical part is something that could cause a significant problem or risk to the facility, equipment, or injury to personnel, when worn, damaged, abused, misused, or improperly applied. Supervisors or work center process enhancement teams (PETs) should be responsible for maintaining inventories of all critical parts, tools, and equipment used on their jobs. They must compile the inventory and identify all critical parts/items by using key staff and existing inventories such as those used by accounting. In addition they should evaluate the use and maintenance of tools and equipment to ensure that the right tools are being used for the right job and that they have been maintained in accordance with manufacturer or process design specifications.

Useful Tools for Developing Inventory

- Quality records, including process flow diagrams
- Risk assessment records
- Maintenance records
- Operators' manuals
- Training manuals
- Job procedures

Supervisors must identify the ~20% of the parts or pieces of equipment that would most likely result in a significant problem if they fail and record this information for future follow-up. Systems safety, process reviews, and risk assessments are critical tools used to evaluate new or existing processes and to objectively determine the readiness of the facility to perform safely and in accordance with customer specifications. Management must also be careful about which risk assessment tools are used. Many risks can be effectively evaluated by using qualitative risk assessment tools versus the more complex and time-consuming fault tree analysis and quantitative risk assessments.

This observation is not an implication that these system safety tools are not valuable. They have been proven many times by the National Aeronautics Space Administration and the Department of Energy to be very effective. However, efficient management of resources dictates that less time-consuming and resource-driven tools can be useful while forming a foundation for more complex risk assessments if needed. The qualitative risk assessment process discussed in Chapter 5 will be more than adequate for most jobs. In this sense, qualitative risk assessment processes can be valuable to sort out the higher risks for Failure Mode and Effects Analysis (FEMA) or Management Oversight Risk Tree analysis (MORT). It is recommended that you develop inventories of critical processes, equipment, and tools, and that you perform initial risk ranking of them by using a qualitative risk assessment process.

PERFORMANCE OR BEHAVIORAL SAFETY MANAGEMENT

(See the references in Appendix I for more information on these measurement and performance management systems.)

Leaders in this field such as Behavioral Safety Technology (BST), Liberty Mutual, Dan Peterson, Scott Geller, and Aubrey Daniels all have similar approaches to executing these systems. The primary differences are in the level of systems safety used to eliminate substandard working conditions. As mentioned earlier, it is the existence of these conditions that often induces substandard behavior or performance from employees.

After a short career with OSHA, I spent most of my early years working at Mare Island Naval Shipyard, a 13,000-person shipyard. Labor relations at the shipyard were marginal, and the union did not trust the former safety program. Costs were also higher than other public and private shipyards. We surveyed employees' perceptions of safety implementation at the facility and their attitudes toward management at the shipyard. Employees had a low opinion of management. The unions specifically labeled this perspective as "negative management".

Our department heads were convinced that it was time to change. The shipyard was 125 years old and saddled with hierarchical and military-type management systems. The rates of accident frequency and severity were higher than average, and worker's compensation costs were escalating in spite of increased OSHA compliance-type inspections and auditing activities. The same problems I saw in EH&S were reflected in our production costs and in difficulties maintaining our schedules. Union trade protection activities and their issues with management also contributed significantly to increased costs. These problems ultimately led to additional management studies and the creation of what we called our management enhancement committee.

I chaired the committee of 30 senior and mid-level managers and employee representatives who evaluated the shipyard management style, structure, and supervisor-management development process. Tom Peter's book *In Search of Excellence* was popular at that time and I was sent to meet with each of the companies discussed in the book to review their management development and quality management systems.

We then drafted the shipyard's management philosophy and prepared policies on supervisor and management development. We also improved procedure development by integrating EH&S requirements into quality and process control documents, making them an integral part of the work process.

To assist the shop's focus on the critical performance elements of their jobs we facilitated work center safety–process enhancement groups. The goal of these work groups was to identify both safety and process risks and design requirements to prevent rework or quality issues. These elements were critical to their jobs with potential to impact schedules and budgets.

The identified critical process elements were then used to create job-task observation check sheets that supervisors could use as a "memory jogger" while performing job observations. Within one year we realized a significant reduction in injury frequency and severity rates while the shops started reducing costs and improving their productivity.

The pilot loss control program incorporated systems safety, process, and quality management with the shop's safety programs. Communications and the level of dialogue between the unions and management significantly increased. Productivity and safety also improved. The program was very effective as we integrated with the shipyard's quality management program and provided supervisors with the tools to identify the multiple root causes of accidents and process failures. The program also had some aspects of today's performance management or behavioral safety programs.

One problem with behavioral management/safety is its name. Many think of it as either behavioral modification, fixing the broken person, or a "Gotcha"-type management tool — a version of old industrial engineering time studies linked with either discipline, reengineering or down-sizing. The reason for briefly describing the process at Mare Island is because the pilot shops successfully integrated quality and EH&S. The job observation check sheets that were developed for them were very effective; they focused the team and, in particular, the supervisor on critical job performance behaviors. We did not try to analyze the antecedents or consequences of behavior that may have improved our process. However, we did modify deficient procedures and the behavior we expected to accomplish tasks safely and according to job specifications. The shipyard pilot involved work center supervisors and employees. Work center teams were facilitated to help identify and focus on critical job behaviors. We then developed supervisor critical job performance check sheets and asked them to use the check sheet as a daily or process start-up "memory jogger". One advantage we had was that nuclear shipyard supervisors were already familiar with written job or process procedures, including signed check-off sheets.

Supervisors were also asked to formerly inspect for process nonconformances and to document this activity at least once a week. Supervisors were expected to correct nonconformances as they were observed and to recognize and reinforce positive activities. In addition, the need for control or design support had to be identified with a proposed target date for correction and, to the extent possible, identification of interim controls. In one year the six participating shops significantly reduced their losses while experiencing increases in productivity and reductions in rework.

Understanding the definition of behavioral management or behavioral safety is essential for process management. What behaviors and associated process design tools enable tasks to be performed safely and in conformance with process quality or safety requirements? How do you consistently prompt employees to perform the desired behavior, and how do you reinforce this behavior or effectively correct nonconformance? What are the antecedents or consequences for desired behavior, and what do you need to do to change the antecedents-consequences of undesired behavior?

Because of the complexity of work processes and because both financial and process failure reports are retrospective, managers oftentimes cannot prevent accidents or process nonconformance. They are basically driving forward looking in the rear view mirror, not able to see problems in time to prevent them. Managers need prospective measurements such as these provided by performance management or behavioral safety programs. These tools measure performance as it occurs, correcting nonconformance and reinforcing desired behavior. The measurements are frequent, often weekly, and based on job observations performed by trained peers and supervisors.

Human error is an important contributing factor to process nonconformance and accidents. However, in most instances it's the last domino in a series of process inputs (see Figure 1) that generated the incident. Human error was one factor of many and most likely the human behavior was the result of sub-standard conditions that are part of the management system. The most important contributing factor is oftentimes poor communications, which is evident in outdated procedures and inadequate or untrained supervisor communication.

In addition, production demands may require employees to remove barriers or to deviate from procedures, therefore, building substandard conditions into the modified (high production) work procedure. To ensure process conformance and prevent injuries, you must identify the antecedents or motivators for the undesired behavior especially incompatible goals.

At the end of the day, communication breakdowns throughout the organization and process are what cause nonconformances.

One of the problems with safety data is that the number of reported injuries is generally a small percentage of other oftentimes undetected nonconformances. Therefore, another reason for integrating management systems is so that more complete documentation of nonconformances are identified. This refinement will increase your ability to determine the special causes or multiple root causes of process failures. By having more complete data you can differentiate from random causes and statistically special or systemic causes. This ability will also help you prevent your organization from reacting to the clustering of random events occuring close together and causing management to infer that there are "special causes" or systems design/implementation failures causing the nonconformance.

Because not all nonconforming incidents are recognized, organizations also tend to reward substandard performance while failing to recognize and reward process improvement. Again this tendency will be caused by statistical randomness and clustering of unwanted events. In addition, the problem with just tracking lost-time accidents is that companies with very low incident rates do not have adequate data to help them distinguish between random and statistically significant events or special causes.

Obtaining accurate and meaningful data in developing countries or in countries with social systems for medical care or employer's liability will be another problem for multinational companies. The available data will often times be grossly inadequate.

Safety data for most companies are incomplete at best, thus requiring a better tool that documents percent of nonconformance with critical tasks and measures job performance. Without better measurements and data collection, companies that have performed well will miss statistically significant events and, in the case of safety, slip into random nonconformance or at best find it difficult to cost-effectively focus attention on further loss reduction.

Common sense should tell you that for serious process nonconformance it is even less likely that the blame would be employee carelessness. A more appropriate question would be, What where the multiple causes at the process input stage that caused the accident or process nonconformance and what were the substandard conditions?

Processes are oftentimes complex as can be illustrated by fault tree analysis and process flow diagrams. Serious process failures or accidents would require a complex plan on the part of the careless employee to cause the number of parallel tasks and sequential steps so that events come together and cause the accident or process nonconformance. As I mentioned earlier, the employee in most cases is the last causal factor generating the undesired event. In most incidents, more than 50% of the causes are not recognized and therefore are not corrected, which is especially the case with less serious problems that tend to be precursors to disasters.

The effects of uncontrolled processes inputs (system root causes) increase the frequency of nonconformances and severity of instances as the process becomes less stable. Organizational control is possible if the nonconforming behaviors and the substandard conditions causing them are frequently observed/measured and if the desired behavior and conditions are reinforced in an environment of continuous improvement.

The best way to prevent nonconformance is by not blaming employees and sending them to training but by providing then with the time, tools, and resources needed to do the job correctly the first time (see Fig. 1). It's interesting that most work center budgets and the managers responsible for them focus on fixed cost management rather than controlling variable costs by improving planned/defined procedures.

Behavioral safety or performance systems are not quick fixes and to be successful they require a commitment to ensure that policies, procedures, training, tools, and equipment are improved before starting the observation/measurement process. Performance management establishes a feedback loop to monitor conditions and resulting employee behavior before the accident or rework occurs.

Employees, supervisors, and the engineers involved with the work process are needed to develop inventories of critical tasks. There is some debate who should conduct the observations, and I am inclined to suggest that both supervisors and employees participate to ensure that observation check sheets are designed that focus on the key steps in the process. Performance management and behavioral safety provide routine intervention and coaching for critical job–tasks, reinforcing the desired process output. There must, however, also be a commitment to ensure that the system has been well-designed and implemented in the first place.

The key to behavioral management is in using the same statistical methods as quality systems SPC. Process flow diagrams are necessary to help define the process as it was designed or intended to be performed and to identify where known or potential nonconformance exists. Liberty Mutual has developed a very useful tool called Risk Safe 98 for performing qualitative risk assessments. This software assists you to rank risks and identify known potential risks of nonconformance. It also produces process flow diagrams, including documenting potential or known nonconformances on the model.

Behavioral observations result in sample data and create the opportunity to measure critical performance. These data in turn give the managing supervisor the opportunity to recognize and reinforce desired performance and to correct undesired nonconformances. The information also allows the work center team to identify the need for a systems review or redesign of the process to improve quality, reduce costs, or reduce the potential for EH&S losses.

Behavioral-based programs are good and can generate significant positive improvements in the process performance and reductions in losses. However, the fundamental process design and applied resources must meet quality process performance requirements, and the system must be implemented in a climate of trust and integrity. Management and the PET must work together to understand and solve problems. Good communications are essential if the process is to succeed. Improvements in process performance will not be realized in a negative non-trusting environment or where management is unwilling to be involved and instead relinquishes its responsibility to committees.

Behavioral-based programs can be a disaster if they are not executed appropriately. Organizational or cultural readiness must be confirmed and understood prior to committing significant planning resources for the execution of a behavioral-based system. If the multiple root causes of losses have been identified and corrective actions implemented, significant loss reduction can be realized.

A system to evaluate on-going performance and to reinforce desired behavior or correct undesired behaviors should be implemented when necessary controls are in place and the organization understands and confirms its own cultural readiness for the process.

INSPECTION AND EVALUATION (MONITORING)

Organizations should establish and maintain documented inspection procedures for routine process QEH&S inspection and evaluations to verify compliance and/or conformance to specified organizational QEH&S requirements and work procedures. Whenever and wherever the potential for illnesses, injuries, rework, or environmental liabilities exists, organizations should develop and document a systematic investigation procedure. The scope of inspections or evaluations should consider degree of process risk.

Routine inspections and evaluations of the working conditions should be conducted at a frequency warranted by the severity and probability of hazards or process risk associated with the job–tasks. These evaluations should proactively anticipate, recognize, and evaluate hazards. They should also confirm that procedures are adequately defined and followed, or they should establish a need for process design change or additional controls.

The required QEH&S inspections and evaluation and the records established for them should document the QEH&S plan effectiveness. Organizations should also have procedures for maintaining these systems records, including audits and inspection records with specified retention periods. In some instances, processes may have detailed check sheets, including preparer/reviewer sign-off or supervisor vital performance or behavioral-based observation check sheets. The retention of these records may also be required by some regulatory agencies such as OSHA or EPA.

Inspection and evaluation reports should proactively prioritize corrective actions based on the severity and probability of identified hazards and the cost-effectiveness of recommended corrective/preventive actions. Initial baseline surveys with periodic follow-up must be established. Critical behavior or processes check sheets can be very valuable communication tools in helping supervisors reinforce desired work/process behaviors or to identify process nonconformance.

In the event that processes or devices do not comply or conform to applicable standards, guidelines, or regulations, effective interim control measures should be implemented until a state of compliance or conformance is achieved. Processes or equipment found to be in conformance should be periodically re-evaluated to ensure continuing process conformance and to reinforce the behaviors of those employees executing the conforming process.

In some cases, medical surveillance of workers may be a statutory requirement or a desirable feature of workplace hazard evaluation efforts. When medical surveillance is implemented, it should be developed and supervised in collaboration with qualified individuals such as an occupational health nurse or a board certified occupational physician. Medical surveillance programs will vary depending on organizational activities. Some examples of medical surveillance programs include the following.

- Pre-placement examinations
- Return-to-work examinations
- Biological monitoring
- Periodic fitness-for-duty examinations
- Illness and injury treatment or interventions

- Basic health promotion activities
- Respiratory protection and hearing conservation programs
- Occupational and physical rehabilitation
- Substance abuse testing

To ensure that process control/site safety is maintained all supervisors should be expected to evaluate the work site by using check sheets or task observation sheets that are specific to the work center or job site processes. The formal inspection process requires that risk be identified and documented, including pictures if possible. Target dates for implementing and documenting interim and long-term corrective/preventive actions must be established. This step is an excellent opportunity to recognize and reinforce conforming process behaviors.

Supervisors must identify the critical 20% of the tasks, tools, materials, or equipment that would most likely result in a significant problem if they failed. This information should be recorded on inventory assessment forms, such as the job–task analysis worksheet in this chapter (Fig. 26). Supervisors are also expected to inspect new work process to ensure that potential process failures or unsafe work conditions (equipment, materials, or tools) are identified and corrected before the job starts.

Measuring Techniques

The following are examples of things that should be done to evaluate or measure QEH&S performance. Measurements can include systematic workplace inspections by using checklists to determine the readiness or appropriateness of tools, equipment, maintenance facilities, material usage, procurement procedures, training, and communications.

Inspections, job–task observations and analysis, monitoring should account for the following.

- Inspections of specific machinery or facilities to verify that safety-related parts are fitted and in good condition
- QEH&S sampling—examining specific aspects of QEH&S procedural performance
- Environmental sampling—measuring exposure to chemical, biological, or physical agents (e.g., noise, dusts, X-rays) as compared with recognized standards
- Behavior sampling—assessing workers' process conformance to identify unsafe work practices that might require correction or to reinforce desired behavior
- Analysis/review of documentation and records
- Benchmarking against best practices, nonconformance rates, and the average cost of nonconforming events (severity) as cost per unit of production

Hazard Inspections Guidelines/Audit Attributes

Inspections should be conducted jointly by management and employee representatives who have been trained in hazard identification, audit, and job/task observation procedures. Inspections or job observation criteria should include input from the personnel performing the observed job to ensure cooperation and the completeness of the process.

- Inspection reports should be forwarded to senior management and the QEH&S committee(s) or PETs as appropriate for consideration in the management review process and for budgetary or resource allocation purposes
- Corrective/preventive actions should be monitored to ensure their effective implementation

An organization should measure, monitor, and evaluate its QEH&S performance to identify the need for additional corrective/preventative actions. These assessments should include the identification of the multiple root causes of nonconformances such as accidents or process failure. There should also be a process for sharing the lessons learned from these assessments with other parts of the facility or organization potentially at risk from similar operations.

Baseline (benchmark) audits must include the following.

a. Review of the current status or level of management systems policy and procedure implementation (Gap analysis)
b. Review of compliance with legal requirements (legislation)
c. Comparison of existing management system framework with industry standards (best practices and Gap analysis)
d. Identification of significant process risks
e. Review of past experience (statistics), previous audits, quality, rework, customer complaints, accident trends, environmental liabilities or releases, and permit compliance
f. Review of current resource allocation (money, time, equipment, personnel, etc.)

Program Audits

Organizations must have a consistent system for providing quality products and services for their customers. Your scopes of work, costs, and deliverables must be consistently executed to meet specifications. The purpose of an audit is to verify that each facility has a documented QEH&S management process to ensure process conformance and that the supervisors and employees understand and executes the process. Auditors will evaluate an organization's physical evidence of the QEH&S process as documented by reviewing these management systems and interviewing employees and inspecting operations for conformance. Audits should include the following.

- Documentation of the customers requirements
- Design and execution of the product/service delivery in accordance with best practices and customer requirements
- Review of the deliverable product/services to ensure conformance to requirements
- Error identification, corrective/preventive action, training, and customer follow-up

In addition, the administrative process should be evaluated for preparing contracts and work orders, customer product or service delivery, contract file management, and contract process integrity. The goal of the quality management system is to help organizations enhance productivity and profitability while ensuring customer satisfaction by reducing QEH&S risks. Quality for purposes of the audit is defined as organizational structure, procedures, processes, and resources needed to implement quality management and the delivery of customer services/products as requested safely, on time, and within or below budget.

Audits are necessary to verify processes for ensuring quality delivery of services. Quality assurance and QEH&S management are all part of the planned activities implemented within the system and demonstrated to ensure that QEH&S requirements are achieved. We include within the elements of the QEH&S control system requirements such as reviewing (inspecting) customer requests for services or product specifications before delivery (validation and verification).

ISO 10011's guidelines for auditing define a quality audit as a systematic independent examination to determine if quality activities and related results comply with planned arrangements and whether these arrangements are implemented effectively and are suitable to achieve objectives. Auditors who do not have direct responsibility for the areas being audited must perform quality audits. One purpose of audits is to evaluate the need for improvement or corrective/preventive actions.

The QEH&S system consists of the organizations QEH&S manual, procedures, policies, and specifications or work instructions. QEH&S systems review include assessments of the following.

- Normal work planning and management control systems
- Progressive quality improvement processes
- Defined and documented management system elements
- Objective evidence of planned process controls
- Periodic senior management review and assessment procedures, including documentation
- Records maintenance

Audit Process Discussion. Management must be able to describe and verify through physical documentation the following program elements. In addition, supervisors, employees, and administrative staff should be interviewed to document their understanding of the QEH&S process and their respective roles within the process. Table 14 outlines some of the documentation required by ISO 9000.

Procedures for planning and implementing the management system and for benchmarking or verifying the effectiveness of the facilities progress along the path to continual improvement should be documented. Part of this process will be conducting internal and external audits. Internal audits should be scheduled every six months to a year depending on the magnitude of the risk and should be more frequent during the early stages of implementing a management system.

TABLE 14. ISO 9000 Requirements for Procedures and Records

Requirements for Documented Procedures		Requirements for Records	
4.3.1	General contract review	4.1.3	Management review (quality system)
4.4.1	General design control procedures	4.2.3	(h) Identification and preparation of quality records (see 4.16)
4.5.1	General document and data control	4.3.4	Records (contract review)
4.6.1	General purchasing	4.4.6	Design review
4.7.1	Control of customer supplied product	4.4.7	Design verification
4.8	Product identification and traceability	4.6.2	(c) Establish and maintain quality records of acceptable subcontractor (see 4.16)
4.9	(a) Process control: documented procedures defining the manner of production, installation, and servicing where the absence of such procedures could adversely affect quality	4.7	Control of customer-supplied products
4.10.1	General inspection and testing	4.8	Product identification and traceability
4.11.1	General control of inspection, measuring, and test equipment	4.11.1	General control of inspection, measuring, and test equipment
4.13.1	General control of nonconforming product	4.11.2	(e) Control procedures: maintain calibration records for inspection, measuring, and test equipment
4.14.1	General corrective and preventive action	4.9	Process control
4.15.1	General handling, storage, packaging, preservation and delivery	4.10.2.3	Receiving inspection and testing where incoming product is released for urgent production purposes prior to verification shall be positively identified and recorded
4.16	Control of quality records	4.13.2	Review and disposition of nonconforming product
4.17	Internal quality audits	4.17	Internal quality audits
4.18	Training	4.18	Training
4.19	Servicing	4.14.2	(b) Corrective action
4.20.2	Statistical techniques, procedures		

External or third-party audits should be performed at least every two to three years or possibly more frequently if you are preparing for ISO certification. Audits are also a great opportunity for cross-facility or peer-review audits, which create discipline in the management system and provide a tremendous learning opportunity for participants. Auditors should verify through interviews and review of documents the effectiveness of system execution as part of their audit documentation process. The results of audits should be recorded and communicated to management having direct responsibility for the audited activity. Auditors should also get concurrence from the audited manager on nonconformance findings and recommended corrective/preventive actions, including targeted corrective action due dates. Deficiencies documented during the audit should be prioritized for corrective/preventive action, and follow-up audits should be scheduled to verify the effectiveness of the planned actions.

Guidelines for Conducting Audits

Scheduling and Pre-audit planning

- *Audited facility*: Pre-audit activities can include explaining the purpose of the audit to the managers, supervisors, and staff of the facility and arranging for the required records and staff support to complete the audit.
- *Auditor*:

  Plan the audit and brief the audit team.

  Conduct audit, including management opening and closing conferences and daily progress meetings with audit team and office staff.

  Obtain concurrence on nonconformance corrective actions.
- *Audit plan*: The audit plan is for the most part designed to force the auditor to evaluate existing systems and processes. Simple "yes" or "no" responses are not adequate. Responses generally require verification and a documented description of the process.

Definition of Major–Minor Deficiencies

- Historically most deficiencies occur with document and service delivery design control, and almost 50% of deficiencies occur with process control. Nonconformance findings must be supported by objective evidence and observations of fact. Nonconformance reports should include the following.
- Identification of where the nonconformance was observed and the facts that describe and document it
- Referencing the standard, policy, or regulatory clause being evaluated for conformance, specific to the standard's sub-clauses.
- Factual data (As an example, 3 out 10 control procedures reviewed were not in conformance with requirements), referencing the specific standard

or SOP, describing the exact nonconformance, verifying it with objective evidence, and pictures

- *Minor Deficiencies:* A single observed lapse in a procedure
- *Major Nonconformance:* The absence of a required procedure or the total breakdown of a procedure

Auditor Qualifications (Training, Experience, Certification)

- Auditors must be experienced management system auditors or have documented training and experience performing similar management audits.

Frequency of Audits and Schedule for Audits

- Each facility should conduct internal audits at least once a year. Independent third-party audits are recommended once every two years.

Audit Process and Lead Auditor Role

- Lead auditors are responsible for objectively communicating and clarifying audit requirements with the manager of the facility being audited. They are also responsible for documenting their observations and reporting audit results, including obtaining concurrence with nonconformance reports and recommended corrective actions.

The lead auditor should plan and schedule audits and is responsible for conducting all management meetings with audited facilities. In addition the lead auditor is responsible for the coordination of the audit report preparation with input from other audit team members, including the development of nonconformance corrective action plans.

Review-Concurrence with Nonconformances. All nonconformance findings should be reviewed with the manager of the audited facility, including obtaining concurrence on the determination of nonconformance multiple root causes, recommended corrective actions, and target dates for completing actions.

Recommended Audit Report Formats

- Executive summary
- Nonconformance report forms and recommended corrective actions
- Miscellaneous documentation, evidence supporting nonconformance findings

Control Procedures for Testing and Evaluation

Organizations should determine required measurements if sample collection or testing is performed and accuracy is required. Selection must be made of QEH&S inspection, measuring, and evaluation systems tools/equipment that take into consideration the required specificity, sensitivity, accuracy, and precision for the sample or test equipment. In addition, there should be procedures for the use and calibration of the equipment, including documentation and training of staff to use the sample procedures and instruments.

If testing equipment is used in the QEH&S process there must be defined procedures for the maintenance and calibration of the evaluation equipment. Calibration records must be maintained for sampling equipment, in accordance with operating/maintenance requirements and as defined in the QEH&S manual and/or documented procedures. Workplace environmental monitoring should be conducted in accordance with regulatory requirements, facility, or industry best practices. Records of monitoring results must be maintained, in accordance with corporate policy and in some cases regulatory requirements.

Proactive and Reactive Monitoring

An organization's performance management system should incorporate both proactive and reactive monitoring. Proactive monitoring could include audits, inspections, or job observations such as behavioral safety or performance monitoring and/or job/task observations. Reactive monitoring should be used to investigate, analyze, and record QEH&S management system failures—including accidents and near misses (nonconformances). The goal of all investigations will be to determine the multiple root causes of the process failure.

Organizations must establish and maintain documented procedures for routine Quality and EH&S inspection and evaluation activities to verify conformance with specified organizational or legislative requirements. The required quality and EH&S records should be defined in the management system manual or other documented procedures, including specified retention policies. In some instances, processes may have detailed check sheets with sign-off or supervisor critical performance or behavioral-based observation check sheets.

Inspections and evaluations should proactively result in prioritized corrective/preventive actions based on the severity and probability of identified risks. Initial baseline surveys with periodic follow-up inspections must be implemented. Critical behavior or process checks sheets can be very valuable to reinforce desired work process behaviors or to identify process nonconformance. In the event that a process does not comply or conform to applicable standards or guidelines, effective interim control measures should be implemented until conformance is achieved. This requirement should also include identifying the antecedents or consequences of undesired behavior and changing them and reinforcing desired behaviors. Processes or equipment found to be in conformance must be periodically re-evaluated for compliance, especially if there have been other process or production changes.

Routine inspection and evaluation of the working conditions should be conducted at a frequency warranted by the severity and probability of hazards or process risk associated with the job–tasks. These evaluations should proactively anticipate, recognize, evaluate, and control risk. The scope of assessment should consider the degree of process risk and confirm that procedures are adequately defined and followed.

JOB–TASK OBSERVATIONS AND ANALYSIS

Many programs are designed to reward or recognize employees who haven't incurred injuries. More successful and permanent programs zero-in on what people do (or don't do) to cause injuries or process nonconformance. In most cases substandard performance is the direct result of substandard conditions. Until you identify these specific nonconforming conditions, process improvement will be difficult.

The work or process conditions creating problems in safety, quality, and production must be corrected to allow the organization to make significant progress with the continuous improvement process. Effective observation and task analysis are critical if systemic causes of nonconformance are to be identified and processes successfully changed. Frequent reinforcement of desired behavior must be initiated to effectively reinforce the desired behavioral outcome.

Process evaluation and risk assessments are essential elements of the system with a correspondingly equal portion of effort devoted to training management and supervisors to use these critical tools. Quality improvement or safety committees should develop a list of the facility's critical tasks, risk assessments, and control prioritization (risk ranking).

Critical tasks are those that are an essential and routine element of the production process. Also, these tasks represent 80% of the facility's process failures, incidents, rework, or accidents. It is necessary for the management team to recognize these tasks and to systematically perform risk assessments and job-task observations by using check sheets or observation tools. It is also important to understand how to perform task analysis with multiple root cause analysis to detect the causes of process failures or accidents.

Job–task analysis includes risk prioritization by using assessment processes and critical inventories of tasks, materials, parts, and equipment. The planned job observation is a tool to make managers aware of current practices in relationship to the defined or expected process. Job observation is a technique to ensure that tasks are performed efficiently and to standards. While making an observation, the supervisor or observer validates employee training, job procedures, adequacy of equipment, and use of correct materials and training.

An observation that reveals proper performance gives an opportunity for behavior reinforcement. An observation of substandard performance gives the supervisor or observer the opportunity to correct identified training, motivation, equipment, materials or maintenance problems that require corrective/preventive actions. An observation of superior performance can also identify potential improvements in procedures or practices. This analysis could be used to share lessons learned, improve safety/quality, reduce costs, and provide management with the opportunity to recognize and reinforce good performance.

Performance of Job–Task Analysis and/or Systems Safety Reviews

It is recommend that you develop a system to execute a process for periodic in-house inspections (checklists) by using supervisor or employee observers performing the periodic observations. However, this recommendation, if adopted, means that you must be committed to correcting the work environment and the substandard conditions in the system or work environment. Until you have identified the specific environmental conditions or the systems influencing the behaviors of people, these conditions will continue to create problems in safety and quality. Organizations will find it difficult to make significant progress with performance improvement or to move very far in creating a safer workplace or improve process quality until the substandard conditions in the system are identified, eliminated, or controlled.

Effective observation and task analysis are critical if the process is to be successfully changed. These aspects allow you to identify the causes of substandard condition processes and to correct them. This process improvement in turn has a positive influence on the behaviors of employees reducing the frequency and severity of substandard performance. Substandard conditions, in addition to being the root cause of accidents, cause people to change procedures, tools, and equipment, which are oftentimes bandages at best and ultimately cause more serious injuries or quality problems.

Job–Task Analysis, Including Reviews of Existing Process Control Instructions

Work center supervisors and observers should coordinate facilitated reviews of high-risk operations using some of the tools listed below, and in Chapter 5 to assess the QEH&S risk, including the risk of product-service nonconformance. When risks are identified the assessment team should work to develop process improvement strategies. These controls could include training programs or the design of new barriers between the employee and the risk exposure. Possible risk assessment tools can include the ones listed below.

- Change analysis
- MORT
- Barrier analysis
- Workplace risk assessment control (WRAC)
- Process work flow (e.g., fish bone) diagrams

Process Management Considerations

- Are the tools, personal protective equipment, procedures, and housekeeping adequate for the job?
- Have standards been established to ensure that tools are maintained in good condition and that procedures are current and reviewed frequently?

Job–Task Analysis Procedures

The logical first step toward process conformance is to develop written job procedures or SOPs. These procedures teach how to do a job safely and with maximum efficiency. They also reduce process risk and improve quality. A job or a task is a segment of work, a sequence of steps a person takes to complete a work assignment.

The uses of work site check sheets or task analysis are the primary tools for evaluating work procedures/controls and employee conformance with procedural controls. Procedures and practices are disciplined ways to complete work properly. The procedures state specific things that must be completed for the job to be done correctly and safely. SOPs state general methods to prevent process nonconformance and accidents. These tools can create a foundation for developing behavioral safety or performance management observation check sheets.

Effective observation is a learned skill, requiring training, allocation of the manager's staff and adequate time for co-workers to effectively evaluate the work process.

Significant considerations:

- Are standardized process control/procedures needed or used?
- Are critical jobs selected and their procedures reviewed on a regular basis?
- Are standard process control procedures used for employee indoctrination and process training?

Job–task procedural controls include safety, environmental management, production, and quality procedures. The best process control instructions document required work procedures, including those for QEH&S, in a single document. To be effective the observation process must ensure that all occupations or trades are identified and included in the evaluation-observation process. The tasks that are performed by each group must then be inventoried and ranked according to degree of risk. Task analysis is also used as a practical training tool to recognize potential risks or inefficient work methods to ensure consistency of work task standards and to involve line supervision in safety and quality control.

Seeing Versus Observing

Seeing is physical, involving muscles and nerves, and is prone to blind spots or scatomas caused by the observer's own expectations, assumptions, and biases. Observing is deeper; it requires understanding, interpretation, and evaluating what is seen. It assumes that the observer understands there are antecedents to behavior or process performance. Either a system is not adequately designed or executed (systems safety problem) or other rewards and consequences are forcing the employee to deviate from the performance standard.

Observation Tips. Effective observation is essential for successful job/task analysis and behavioral-based safety and performance management programs. Stop and record observations, take the time to watch and think about what is happening. Be inquisitive! Use all of your senses by looking, listening, feeling, and smelling for changes in the process or work environment. Think about how the job could be done better and solicit ideas from employees. Don't make employees wrong; evaluate why the process failed or why the observed behavior was necessary from the employees' perspective. Encourage all levels of the organization to think of how to do the job more safely and with less rework.

Accidents and process nonconformance usually result from both substandard conditions and acts. However, the work environment, including all of the process inputs identified in Figure 1, tend to cause the substandard performance, especially when communication, including training, procedures, and supervision are inadequate. Supervisors must be alert to perform task observations and take immediate corrective actions to prevent injuries or rework. In addition, it's equally important to recognize and reinforce positive behavior and conditions. Competent persons should identify potential hazards and assess the risks from the work process. Hazard identification and risk assessment procedures should be documented (e.g., hazard inventory, consultants reports, manual handling assessments, noise assessments, etc.). Documentation should also describe the hierarchy of controls (engineering, protective equipment, administrative) in the determination of control measures.

Steps for Observation Success

- Review existing procedures with emphasis on critical problems and loss potential
- Set observation priorities
- Focus your attention on the task
- Reduce distractions

Observation Checks

- Are tools and equipment in safe condition, and are available procedures used correctly?
- Are procedures and tools/equipment maintained, and are they used effectively and in compliance?
- Are housekeeping standards established and maintained, and are they adequate for the job?
- Are guards and barriers being used, including lock-out, tag-out procedures?
- Are positions and actions of people and material handling, contributing to potential risk of injury?
- Are hazardous materials or exposures to other hazardous substances or operations and tools recognized and controlled?

Performance Analysis

- Identify unsatisfactory performance
- Does the individual know that his or her performance is substandard?

Process or Work Center Identification:	Date:		Approvals: ______ Signature/Title Date	
	Completed By:		Other: ______ Signature/Title Date	
CRITICALJOB ACTIVITIES, or PROCESS STEPS	RISK EXPOSURES: Quality or EH&S	QEH&S CHECK Yes	No	RECOMMENDED CONTROLS

MTN Associates

Figure 26 Job-Task Analysis Worksheet.

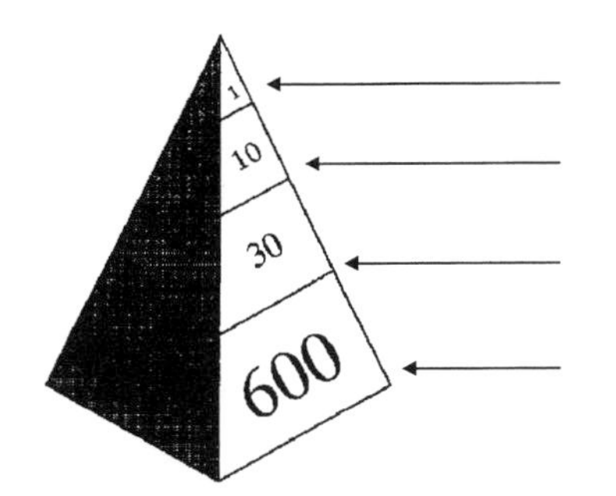

Serious or Major Injury
Includes disabling and serious injuries
(ANZI-Z 16.1, 1967 Revised, Ratio of 1-15)

Minor Injury
Any reported injury less than serious

Property Damage Accidents
All types

Incidents with No Visible Injury
(Near-accidents or close calls)

Figure 27 *Accident Ratio Study.*

- Does the individual know what needs to be done and when?
- Are there obstacles beyond the employee's control?
- Are substandard conditions designed into the process so that the process needs improvement?
- Does the person need training?
- What is the purpose of each task step, and can it be done better?

Benefits of Task Analysis

- Provides information for establishing procedures that can be used for job/task orientation and training observation checklists
- Identifies safety or production quality control problems
- Allows you to eliminate potential problems or minimize them through process controls, including the improvement of process design and procedural controls
- Provides a supervisory check sheet, training guide, and performance management task observation tool
- Assists with accident investigations
- Allows for employee participation in the job planning process
- Identifies substandard systems and communications causing the nonconforming condition
- Establishes a proactive management measurement tool to reinforce or correct performance

Performance Management

Behavioral safety or performance management can work only with good well-defined processes, which are not possible if the controls are inadequate or outside the employee's control. The problem with attitude and incentive-based management systems is that they are ineffective in identifying and reinforcing desired behavior. They also oftentimes have a tendency to pit individuals, departments, or regions against each other rather than focusing on customer needs and maximizing the efficiency of the corporation.

Lincoln Electric is the world's leader in effectively recognizing and rewarding desired behaviors in their employees. This commitment is evident in the fact that their incentive program has been in place since 1932, and since 1950 employees have been going home with Christmas bonus checks equal to 90 to 110% of their regionally competitive salaries. (See Appendix III for a brief introduction to Lincoln Electric's bonus system)

Lincoln measures employee output in manufacturing on a daily basis, and all employees are evaluated on output, quality, dependability, and cooperation twice a year. This system sounds very simple, and in many ways it is. However, the antecedents to good process performance are a well-designed production process that is clean and well-organized. Lincoln used "just in time" material inventories before they were popular. More important to note, employees know that the results of their performance are immediate or near term, creating significant team peer pressure and personal financial results.

As an example, the team that originally built a welder that is shipped with a defect will repair or rebuild it on its own time. Employee evaluations result in a score used to calculate individual bonuses, which has historically been very significant, therefore high performance is expected and realized, including employees repairing mistakes on their own time.

Most companies blame workers and maintain very crude cause definitions for process failure. Measurements, if any, tend to be biased or manipulated. The causes of nonconformance are often outside the employee's control and in reality are unfairly perceived as the result of the employee's carelessness. Process conformance barriers exist and are not adequately controlled, managed, or maintained. We must be able to distinguish system results and behavioral results and understand the multiple root (system) causes of process failures or success. Failures to understand the systems approach to process management will tend to result in short term or erratic success at best.

When performing risk assessments, you must prioritize risks, jobs, or tasks for analysis. Systems approaches to process analysis or job–task analysis look at process inputs, tools, and allocated resources.

- Machines, tools and equipment, facilities
- Training
- Procedures
- Maintenance
- Supervision
- Changes in the process
- Communications, including training and supervision

Key: Are the desired process steps or performance behaviors actually under the employee's control? If not you cannot expect consistent process performance! Are the system or process inputs defined in Figure 3 creating substandard conditions, which in turn contribute to substandard performance or behavior from employees?

When evaluating the work process, view it from the perspective of performance management. What are the antecedent and behavioral consequences for the desired performance? Also, analyze the situation from a systems and a behavioral perspective.

- Are the desired behaviors or process steps achievable with the existing resources available to the employee?
- Are there other consequences in competition with the desired performance?
- What are the antecedents and behavioral consequences for the desired actions or tasks?
- How can the process be changed and positively reinforced by changing process inputs or antecedents, reinforcements, or consequences?

Before doing an ABC (A, antecedents, B, behavior, C, consequence) analysis, complete these two tasks.

- Determine severity and frequency (probability) of substandard behavior
- Determine if antecedents (system) are under the employee's control and evaluate whether the existing management system antecedents generate an undesired behavior? If so, why, how, and what can be done to change the antecedents or system reinforcement to generate desired process performance?

Combine ABC analysis with system safety and job–task analysis when assessing critical performance to ensure to the extent possible immediate positive reinforcement of desired performance.

Rule: Management should be constantly looking out for the good, reinforcing it immediately or as soon as possible versus the traditional management by exception or negative management.

Management must also be aware of the Pygmalion effect, a process in which managers, parents, and teachers have the ability to subconsciously create or cause their expectations in people or situations to be realized. This phenomenon has been demonstrated repeatedly in research with both students and employees. Students were randomly selected and placed in a classroom to be tested and the teacher was told that the students were exceptional and would all be high achievers. The control class came from the same pool of students, but the teachers were told not to expect higher-than-average performance. The test group of students under the teacher's positive expectations exceeded the performance of the control group. Likewise, managers must distinguish good external competition and try not to create internal competition. What football team succeeds by its players fighting over who is going to carry the ball or their individual role in a specific play?

Pinpointing and measuring performance can be simpler if process control and systems safety have already generated well-defined process control instructions and feedback for employees and work groups. Feedback should be daily if possible; but weekly is also good and is strongest when it is coupled with reinforcement, for example, by use of observation check sheets.

I hate resorting to sports analogies, however, you look at team sports and consider why they succeed. Reflect on what is different in work situations, and the room for management improvement becomes apparent. Consider the following qualities in sports teams.

- Specific job responsibility
- Teamwork
- Measurable results
- Challenging but attainable goals, immediate feedback
- Positive reinforcement from peers, fans, coaches
- Emphasis on behavior (Daniels, 1989)

Thomas Gilbert in his book *Human Compliance* estimated that the average plant could improve performance by 300%. Although this estimate might seem high, compare it with Phillip Crosby's estimates for rework in manufacturing averaging 25% and the results generated by Lincoln Electric to allow for bonuses equal to 100% of salary. Therefore, no matter whose statistics you believe, you can see that we have much room for improvement.

Behavioral management in the workplace relies on statistical process control. It systematically identifies and reinforces critical process performance factors. In the process it identifies those elements of the existing system that may be causing rework or forcing employees to perform unsafe acts. With employee involvement it identifies critical performance factors and establishes a way to measure them.

Too often we want to measure the negative. In fact some MBA programs promote management by exception, greasing the squeaky wheel, and error measurement or defects versus percent positive conformance measurement. Ideally, performance should be measurable in four categories.

a. Quality
b. Quantity
c. Timeliness
d. Cost

Effective management systems focus on critical job performance factors—those 20% that generate 80% of the desired outcomes if we the Parieto 80–20 principle or Deming's 85:15 ratio of common cause versus special cause of nonconformance.

Feedback is essential for learning and as a reinforcement of desired behavior. When it is associated with related consequences or reinforcement, feedback must provide specific information about the desired behavior and its outcomes and should address performance that is under the control of the individual. Feedback should be easily understood, self-monitored if possible, and graphed, as in the sample observations report chart above.

Line graphs shoving upper and lower confidence limits, like those used for SPC tend to work the best and are easiest to understand. Self-monitoring, when linked to supervisor review and reinforcement by using feedback as the antecedent to reinforcement, is especially powerful when linked with a combination of social and tangible reinforcements providing feedback and reinforcements. The *coaching* of the work team is one of the most important leadership responsibilities of a supervisor.

Graphs of results should also be designed to show positive versus negative results, for example percentage of safe behavior, percentage of products produced without rework. Feedback should also be goal-related, easy to understand, and reviewed weekly with tangible reward systems whenever possible. In general, the horizontal or x-axis should denote time, while the y-axis measures performance.

Goal setting should include appropriate sub-goals if you are a long way from attaining the ultimate goal. Goals should be SMART (specific, measurable, and achievable within a reasonable time). High achievers also tend to be risk-takers, therefore it might be appropriate to establish stretch goals 20% greater than average. In other words, hold the employee responsible for the average goal, but work as if targeting for the stretch.

In any case, the average goal must be reasonably attainable. The stretch goal tends to ensure obtaining the original goal and in many cases exceeding it. However, it is better to set the goals too low and exceed them than it is to frustrate people with unattainable goals.

Behavior Versus Attitude

Behavior can be measured and therefore managed, whereas you cannot measure attitude. However, changes in behaviors can lead to changes in attitude.

Antecedents

Antecedents influence behavior individually and much less powerfully than consequences, which are also significantly influenced by the certainty and timing of the consequence. Positive reinforcement or consequences are more powerful than negative consequences. The strongest consequences are immediate, certain, and positive, whereas the weakest are realized in the future, uncertain, and negative.

Consequences

Traditional QEH&S programs measure the results of consequences from hundreds of near misses and therefore fail to recognize the significance and frequency of the process nonconformance. Behavioral management programs measure and report positive results, moving from blame to understanding and controlling the work process. Traditional programs and measurement/reaction to random effects result in improper reinforcement of behavior.

At Mare Island, the demonstration project worked well because of supervisor and employee involvement in identifying critical process performance. It was also successful because we integrated our approach with the quality enhancement effort.

Systems safety, check sheets, fault tree analysis all help with process definition-understanding. Once the process is adequately defined, you have the opportunity to identify the potential for improvements or where process goals are at risk. The process also identifies the critical tasks that can be observed to ensure and reinforce process conformance or to correct process deviations.

If the process is a high-risk operation then process control check sheets, including signed conformance signatures, may be required. The use of check sheets for high-risk operations is an antecedent to desired performance and creates immediate feedback for desired behaviors by letting the employees know that they have complied with the process controls, and met performance expectations.

Measurement of upstream factors affecting performance (ABC analysis) is critical to understanding behavioral management. Performance management includes the following.

- Requires operational definitions of critical job performance factors, including the preparation of critical behavior inventories
- Requires observation, measurement, and feedback
- Results in facilitated task assessments and improved process control
- Helps organizations realize process sustainability

In most process nonconformance incidents or accidents, employee behavior was the final domino in a chain of events. Therefore, understanding the process and reinforcing the conditions and behaviors critical to successful performance is vital. It's the hundreds of near misses associated with employee behavior that leads to the serious injuries or fatalities. Therefore, it is recommended that facilities develop a percent-safe score and system to provide feedback and improvement. Measurement of upstream factors affecting performance is critical to process enhancement success.

ABC Analysis (A Antecedents, B Behavior, C Consequence)

Step 1

- List antecedents to undesired behavior
- State behavior in observable/measurable terms
- List consequences of undesired behavior

Step 2

- List antecedents to desired behavior
- State behavior in observable/measurable terms, list positive consequences for desired behavior, designing them to be immediate, certain, and positive

Process risks and control alternatives must be prioritized. No company can correct all risks at once. Table 14 is an example or tool that can be used with your qualitative risk assessment process, including ranking of control options. However, note that control option ranking should also include to the extent possible cost–benefit analysis.

Behavioral safety or performance management systems include the process employee and supervisor and observer involvement in the process and system review. It must also include risk assessments and task analysis to control those elements of the process that are not controllable by the employee. In fact, to be successful with performance management systems companies must control process inputs and working conditions, especially procedures and communications so that they do not become antecedents to unwanted behaviors.

Like TQM, it requires process measurement, management, review, and feedback. Its distinction and strength is the heavy reliance on employees to help define risk and the necessary controls. The reason loss source analysis and standard benchmarks like frequency and severity rates are not as effective is because the sample size is too small. They simply identify location and very broad loss categories and are a measurement of the negative consequence in the

TABLE 15. Risk Rank System

	(Qualitative Priority or Significance Ranking)				
Risk Significance (1–4)					*Range*
4	[8–32]	[12–32]	[16–32]	[20–32]	(8–32)
3	[6–24]	[9–24]	[12–24]	[15–24]	(6–24)
2	[4–16]	[6–16]	[8–16]	[10–16]	(4–16)
1	[2–8]	[3–8]	[4–8]	[5–8]	(2–8)
	Probability: Certain-Immediate Future				
	(2–8)	(3–8)	(4–8)	(5–8)	

Occurrence Uncertain/Certain (Ranking: [1–4])
Occurrence Timing, Future/Immediate (Ranking: [1–4])
Add Certainty and Timing Ranges, then multiply by Risk Significance. Establish Low-Moderate-High Relative Risk Ranking.
Risk Ranking for Management Prioritization of Action Planning
High Risk = 25–32
Moderate Risk = 9–24
Low Risk = 1–8

individuals and the antecedents causing the performance outcome. Performance observations take you from a reactive mode to proactive process control. When linked to a well-designed management system it should significantly reduce loses while increasing profitability.

The measurement of upstream performance versus downstream consequences is meaningful because it measures and reinforces critical process antecedents, vs. rework, nonconformance, accidents, etc. that can not be changed. Measurement and reinforcement of the indicators of quality performance and process conformance, including safety, will always be more powerful than measuring consequences, especially if they are negative.

This point is analogous to the example of my father and the assumption that, even though he smoked and had a diet heavy in saturated fats, he was healthy. My father believed in his health because of his physical strength and the lack of easily observable consequences of his lifestyle. He had not yet experienced the pain or debilitation of coronary artery disease or a stroke. Soon, however, his first evidence of poor health (consequence) was a stroke, followed by a heart attack and a second stroke that ultimately killed him. If my father had been measuring the days he had conformed to a healthy diet and cardio-vascular exercise and had viewed these practices, rather than the recognizable symptoms of a stroke or heart disease (consequences), as antecedents to good health, he might be alive today.

How does the management system influence critical behavior? A systems approach to evaluating the existing system is recommended by many safety or management behavioral performance experts before intervening with behavioral enhancement systems. To accomplish this task, facilities conduct what is oftentimes referred to as a climate culture review and process risk assessments. The culture review gives management a reality check on how the various management and employee components of the organization perceive the existing QEH&S programs. The facilitated risk assessments then give work teams a systematic way to help identify where they have process risk and process changes, including process redesign or other engineering changes necessary to control the risk. The climate culture review, initial audits, and risk assessments also create internal benchmarks to focus on basic risk factors (the continuous improvement process vs. focusing on other companies or industry performance standards, which may or may not be significant or even relevant to your facility).

Good management systems establish a process for measuring performance and should include the measurement of the antecedents to successful process execution. Management must then periodically review the proactive antecedents of expected performance and the consequences of traditional frequency and severity measurements of nonconformance to review the effectiveness of their management system's effectiveness. This periodic review and performance results gives senior management the opportunity to be involved in modifying the policy (or the direction that their corporate ship is taking before it crashes on the rocks in a complacent or bureaucratic fog!). The continual improvement process defined

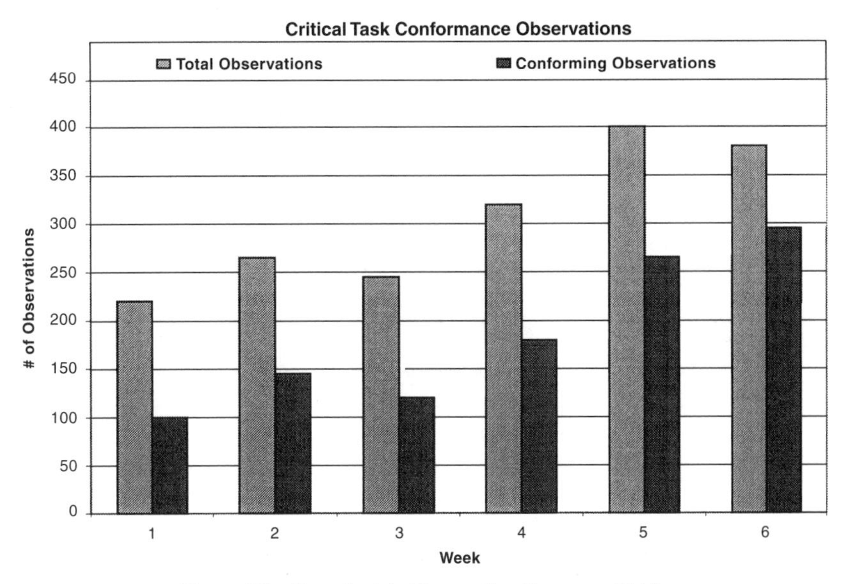

Figure 28 *Sample Job Observation Summary Table.*

by ISO 9000 and any well-designed quality management system require periodic senior management reviews and involvement in the improvement process.

Behavior is a scientific term relating to observable measurable actions. From a management perspective, we want to have SMART goals and objectives — goals that are significant, measurable, or observable; reasonably obtainable; and time-bound. It is difficult to see how the goals of the behavioral manager and the operations manger differ, other than in terms of how they speak or describe their systems.

Employees are oftentimes blamed for poorly designed processes that fail or for their injuries rather than management trying to understand what happened within the system that caused the problems. Management must clarify expected process performance or behavior in psychological terms as action antecedents and consequences. Management must also recognize that they and the system they control generate both good and bad antecedents related to their work activities.

Employees do not go to work planning to hurt themselves — with the exception of fraud cases, which constitute a relatively small percentage of workplace injury claims. However, employees are oftentimes put in a position to take process risks to meet production demands or to accommodate substandard process design or tools. In addition, most accidents are with new employees and older complacent workers.

Figure 29 below clearly illustrates why it is important to evaluate minor incidents, because they are the "red flags", the precursors to more serious consequences or outcomes. If a process is performed under the particular circumstances or variable conditions, personal injury, serious damage, and/or environmental problems will result. Recognizing the process nonconformance

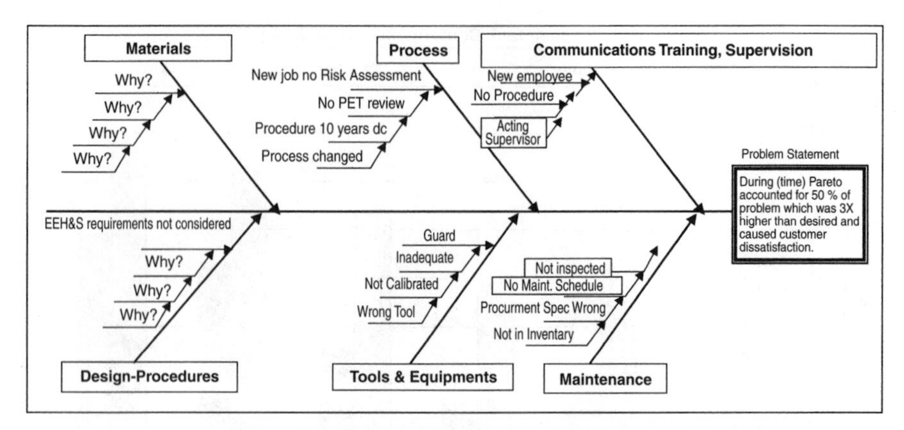

Figure 29 *Ishikawa "Cause and Effect: Diagram.*

and correcting the basic root causes as manifested in a minor incident allow management to prevent injury and quality or environmental risks.

Conclusion: Hundreds of Exposures Precede A Serious Process Nonconformance, Accident, or Fatality. Management systems design risk into the process directly or by failing to recognize and correct risky employee behaviors. Genuine culture change and integrated performance results, real process change, honest communication, and involvement at all levels cannot occur until senior management ensures that the entire management team understands and accepts responsibility for successful process performance.

Management systems must effectively measure upstream process performance to be effective. The key is identifying what to measure and how. In addition, reporting and feedback or follow-up on corrective/preventive actions are necessary. Observers must competently identify physical hazards, perform risk assessments, and identify at-risk behavior, including conducting "What if" scenarios and drafting desired/expected process flow diagrams that clearly identify process deviations and their causes.

Employee involvement via quality circles PETs, and safety committees is critical in correcting processes and risky behavior. Continuous improvement, including behavioral change, creates opportunities for improvement when at-risk behavior or processes are identified and controls or process enhancements are recognized, and implemented. Process enhancement includes the following factors.

1. Problem identification, including multiple root cause and behavioral analysis
2. Brainstorm/facilitate work center ABC and risk analysis
3. Systems engineering, systems process review
4. Identifying potential solutions

5. Testing/evaluating potential solutions
6. Developing implementation or process change action plan
7. Measuring, evaluating, and following-up

Problem-solving groups should rank identified process risks by using experience and observation data, corrective action reports, process nonconformance, or injury data. In addition the results of audits and systems review can be invaluable. Group process change recommendations and commitment of resources should be based on degree of risk (significance), including the financial impact of nonconformances, proposed controls, and amount of exposure (frequency).

Consequences of exposure, including level of certainty and improvement potential, should also be considered in developing a prioritized risk management plan. A simple ranking process would draft a problem statement and design SMART solutions.

Multiply frequency and severity ratings; the greater the number, the more significant the potential risk and therefore the greater the need for resources to eliminate or minimize the risk.

From a quality perspective, consider the many opportunities for improving performance, eliminating redundancy, rework, and injuries through better process design, reengineering, procurement, manufacturing, sales, safety, environmental management, information management, and product distribution. If you think about these opportunities, you can begin to appreciate what some companies like Lincoln Electric have accomplished and the estimated room for improvement cited by Phil Crosby and Thomas Gilbert. In addition, we have companies such as Motorola and General Electric leading the way with Six Sigma programs, which is discussed in more detail later in this chapter.

Behavioral management does not try to fix people or understand why they dislike their mother. It is not behavioral modification. People are generally not broken. In most instances, it is the process that needs improvement. Performance management accepts people the way they are and looks at what motivates performance and what the antecedents and consequences are to behavior. It then focuses on how we reinforce the desired behaviors as quickly as possible after they are done. Some of the problems with most motivational reward programs are that people are rewarded with stuff they don't really want or else the recognition or bonus is not effectively linked to the desired performance and is oftentimes bestowed long after the task was performed.

Behaviors are influenced two ways, by antecedents or by events leading to the desired behavior and consequences. Again, the three components (the ABCs) of behavioral management are Antecedents, Behavior, Consequences (Performance Management, by Aubrey Daniels, 1989). Behavioral management experts analyze performance by using ABC analysis. Antecedents convey information and range from modeling co-worker behavior to forming detailed check sheets for critical jobs. The effectiveness of an antecedent depends on the correlation with the consequences, the lag time between the behavior and consequence, and whether the consequence was positive or negative.

The most powerful form of behavioral management and reinforcement occurs when the consequences are positive, immediate, and certain. Consequences are therefore the most significant tool that management has to increase performance, enhance quality, and reduce accidents. Every behavior has consequences and people do what they do based on their consequence history. Once we understand the consequences people are experiencing as it relates to job–tasks, it is easy to understand not only what they do but what can be done to improve performance.

Four types of consequences can influence behavior.

- Achieve goal
- Avoidance
- Receive unwanted consequences
- Loss of control

The first two reinforcements are the most powerful, whereas the last two punish, which have less influence on behavior. Positive reinforcements are readily available in the form of praise, attention, and various forms of recognition, and they have positive effects on other behaviors and overall attitude.

Problem-Solving Steps

- Identify and define process nonconformances or possible process deviations (flow diagrams work well with this step)
- Identify multiple root causes of process nonconformance in the system
- Hold group-facilitated risk assessments and development of potential control options
- Develop action plan
- Conduct follow-up evaluation
- Develop observable critical performance factors
- Observe and chart critical performance, percentage of safe behavior and or percentage of process conformance

The cause and effect diagram below, when used with other systems safety tools, the critical process input model in Figure 1, or the decision tables in Chapters 4 and 8 can very powerfully assist the PET to define and illustrate the multiple root causes of a process failure. It can also be used more proactively to help understand the potential risk or lacks of readiness for a new process undergoing a risk assessment prior to start up.

The Pareto chart and data below illustrate how percentage and frequency help identify problem areas, which should require a higher priority with regards to resource commitment.

Steps in ABC Analysis

- Describe undesired behaviors
- Describe desired behaviors

Product Returns 98	
Location	Returns
SF	15
SD	25
LA	76
ATL	33
NYC	68
DC	33
Total	250

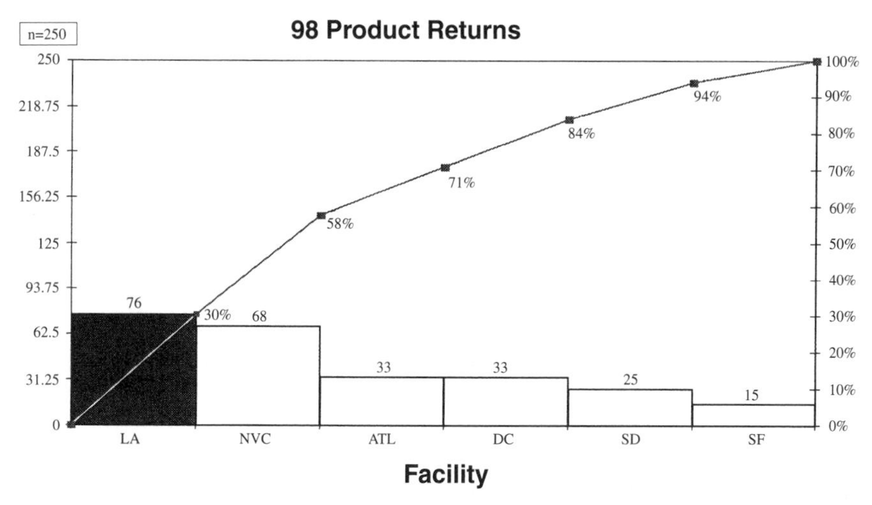

Figure 30 *Pareto Charts Showing Frequency and % Return.*

- Assess the risk or severity of the problem and determine the frequency
- Complete ABC analysis
  - List all possible antecedents and consequences
  - Elevate consequence not relevant to the worker
  - Rate consequence as positive/negative (P/N); immediate/future (I/F); certain/uncertain (C/U)
- Complete analyses for desired behavior
  - List desired behaviors
  - Do step 4 for desired behavior, generating positive, immediate consequences (PICs), and negative,immediate consequences (NICs) as the most powerful tools to reinforce the desired behaviors
- Change/Solution: Generate and reinforce positive, immediate consequences and antecedents to reinforce the desired behavior

ABC analysis is most appropriate for frequently occurring behavior. Training is oftentimes the antecedent for desired behavior (*Performance Management*,

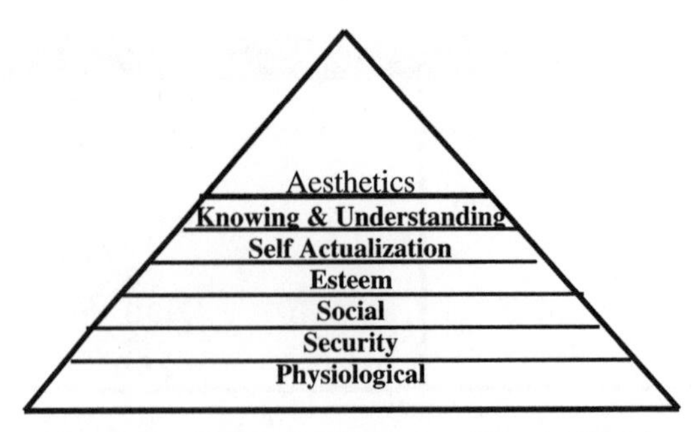

Figure 31 *Maslows, "Hierarchy of Needs' (1943).*

by Aubrey Daniels). One way for supervisors to provide positive immediate consequence is to be visible in the work area, which Tom Peters referred to it as "management by wandering around." Jack Welch, General Electric's CEO has taken this strategy one step further with Six Sigma and will not allow managers to be promoted to executive unless they are trained and show project experience with Six Sigma. The Six Sigma processes, unlike "Management by Wandering Around," are very focused and require hands-on executive management involvement.

If the work group team is involved in the observations, then everyone will know what is expected of them and will participate in providing reinforcement. Negative reinforcement should always be a secondary reinforcement, as it may generate unwanted behavioral side affects. Positive reinforcements have positive side effects and improve morale and team cooperation.

Maslow's Hierarchy of Needs, Figure 32, is useful for understanding positive reinforcement. Money does not provide enough reinforcement for most people because, in most cases, there is not enough and it is not certain that it will be received, and many times it is too far out in the future to be significant. It is also

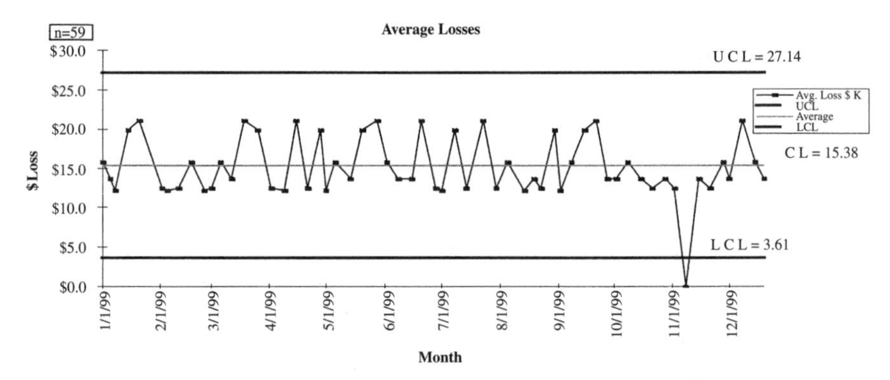

Figure 32 *Sample Chart Showing Potential Upper & Lower Control Limits, Using some Data from Above.*

interesting to note that in an environment of positive reinforcement, people seek out ways to be measured and accountable. Employees working in a culture of positive reinforcement will be more adaptable to change and not as sensitive or defensive about performance-improving suggestions.

Reinforcements follow behaviors and we must identify what they are at both the group and individual level to be effective. Asking or observing employees can identify reinforcements. A reinforcement survey could also be done to identify those things that would be desirable tangible awards. Positive reinforcement should outnumber negative reinforcement.

The best reinforcements are those that are readily available. Tangible reinforcements such as money or gifts may require future recognition and planning, Tangible reinforcements must also be accompanied by social reinforcements, time, and attention from the boss, praise, or other forms of related and relevant recognition.

I once had an environmental consulting company with offices in four states. We adopted Lincoln Electric's bonus system, and every year at Christmas time we had a dinner for each of our offices, which included their spouses or significant others. I gave each employee a Christmas bonus check at the dinner but only after I, as president of the company, called each one to the front of the room to personally recognize each individual for specific contributions to our business that year. I formed this practice from my own personal knowledge of their contributions and from what I considered to be significant.

The staff was recognized in this way so that our associates knew that I was personally aware of the employees' accomplishments. This recognition was given to them in front of their spouses and co-workers. We wanted them to be recognized not just in terms of a business envelope handed to them with a smile or dropped casually in their "in box". Years after we merged with a larger public company, former employees told me that the bonus check dinner was something they would always remember and appreciate about our company.

I also took time at the beginning of the dinner to give everyone an overview of what the company as a team had accomplished over the previous year and to define our goals for the coming year. In addition we shared some thoughts, revelations, and ideas from our annual employee survey. CEC conducted this survey every fall and asked very specific business operations process questions about those things that helped or slowed them down the most. We solicited ideas for improvement and gave the staff the opportunity to support their responses in a narrative.

I found the narrative responses most useful and revealing and shared this insight with the staff. Again, my goal was to reinforce getting their input. The bonus system and employee survey started off slowly, but after we demonstrated the credibility of the process we felt like CEC had 75 partners rather than 75 staff members, with most of them focusing on how we could cut costs and improve our processes and profitability.

Management should be on the look out for opportunities to provide genuine praise. Reinforcement must be as soon as possible after the behavior was performed, and it must be specific to the desired behavior. Recognition should be personal, sincere, and specific to be effective. Don't mix positive and negative reinforcement, or positive reinforcement may be perceived as negative. The best reinforcements give employees the opportunity to identify critical desired behaviors and to monitor or observe them in the workplace while reinforcing the positive behaviors; thereby reinforcement affects the entire group.

Unlike Lincoln Electric's bonus plan, most profit-sharing plans are rarely contingent on individual performance; some generate unwanted inter-organizational competition. Worse yet, the mindless, rotating employee of the month reward systems do little or nothing to reinforce desired behaviors. When punishing or correcting undesired behavior you must be very specific and address the behavior while not making the individual "wrong," or "less than." Just as with positive reinforcement, correct undesired behavior as soon as possible. However, I have found that when I attempt to correct behaviors while still angry the overall process was not as effective. Sometimes it is better to wait for the "smoke to clear."

Again, it is always better to catch someone doing it right and reinforce the positive than it is to correct or punish behavior. Never correct publicly! Both negative and positive reinforcement must be applied consistently. Always ask yourself, "What is it about the system or process that causes the employee's substandard behavior? What are the system substandard conditions at the foundation of the nonconformance?"

Punishment if used too often or incorrectly will generate escape or avoidance behavior and could decrease desired performance. A better alternative is to with hold reinforcement for the undesired behavior and to positively reinforce the desired behaviors. Although this approach may take some time, it will have fewer side effects than negative reinforcements. Reinforcing alternatives can take longer but should have fewer side effects. The more frequently positive reinforcement is used, the faster the desired behavior will become the way.

To be effective in influencing behavior, we must describe performance in detail. This description could include job–task analysis and process flow, including fish bone diagrams or procedural control instructions if they are written precisely and simply. Once the process is adequately described, an inventory of critical job behaviors can be made. This inventory should be specific to the desired performance and should address all critical elements of the process, including safety and environmental requirements. Observable measurable behaviors listed on the inventory sheet must be active and not the outcome of a behavior. Behaviors are active, measurable, and observable, whereas results are static and inactive. We need to measure and understand behaviors and results, but to influence outcomes we want to identify and reinforce those behaviors that create the desired results.

In Japanese quality management systems "Kaizan" means improvement, concerning both management and employees. This philosophy addresses continuous improvement in all aspects of life, business, social, and personal. It covers a set of goals to shape the ultimate desired results. One of the problems with continuous improvement is keeping management focused. The tendency is to devote resources to increasing accident or defect rates and as performance improves, attention is shifted elsewhere. This tactic ultimately causes the defect or accidents to increase again, causing another shift in attention while frustrating the people working for the manager and raising credibility issues for both the manager and the company. Management support and focus must be consistent and committed if continuous improvement is gong to be realized.

Companies realize continuous improvement when they develop methods to track process performance. *Safety, quality, and environmental management are all facets of the same stone.* It does not make sense to focus on one aspect while ignoring the others. Ultimately, we are aiming for better process control. Measurements of the upstream process factors that are predictive of nonconformance are the critical factors that must be observed and monitored for performance improvement. Traditional measurement of accident frequency and severity rates is as useful as declaring my father healthy even though he smoked and had a diet very high in fats simply because he did not have a heart attack or physical evidence of cancer.

As noted in Chapter 2, employees are too often blamed for their accidents rather than identifying the multiple root causes controlled by management for the process failure. At the heart of a behavioral management system is employee management joint identification of critical factors and an on-going effort of employee data collection, generating behavioral data that focuses on upstream critical performance success.

Employees also need to be involved with the identification and removal of process obstacles and to participate in facilitated work center groups to rank and prioritize process risks and the design of necessary process changes or controls to eliminate or reduce the process risk.

QEH&S Shared System or Process Elements

Cost and improvement can be realized through effective process change, reinforcement of desired behaviors, and SPC. This process includes the measurement of objectively defined upstream predictors of desired performance. Process control management must occur upstream; otherwise you are simply measuring results. Deming talks about driving out fear. Most incident investigations fail to identify the multiple root causes of process failures, and instead they look for someone to blame, usually the employee, and many times fail to implement the necessary corrective/preventive actions.

Far too often the recommended fix for the incident is disciplinary action and training for the employee. The injured employee is admonished for being so stupid or careless to cause injury. Blaming is a major source of fear and must be removed from the company's culture if successful process improvement is desired. Injuries and environmental problems come from the management system. The employee's behavior is usually the last domino in a series of process factors that manifest themselves in the form of an injury, rework, and environmental liability.

It is easy to recognize and reinforce desired behavior through peer observation and verbal feedback. Group-facilitated teamwork is also necessary to effectively identify and prioritize risks. The team can select the critical job performance factors to be used in the inventory for job observations. Controls must be evaluated, and barriers to effective performance must be removed by management.

Benchmarking

How do we compare with our historical performance, other work centers, facilities, or competing companies?

- Benchmarking should be done as an indicator of desired performance, establishing accountability, and generating the opportunity to provide feedback for performance improvement
- SPC is necessary because without it managers are rewarded based on random process variation

Frequency and severity rates or insurance risk management loss source analysis fail to identify or measure the critical factors necessary for process conformance. This tendency is because the measurements are of results, and the data sample size is often too small to be significant. It also focuses on the immediate causes of process failure versus the multiple root causes within the management system that attribute to the process failure. Under reporting also tends to compromise the validity of the already statistically insignificant database. Reporting and tracking of all incidents is better but it is still a measurement of results. After collecting data, management should ask if the periodic variation in the accident frequency is due to common cause or special cause. To determine the type of cause, these steps must be taken.

- Plot the data
- Calculate the mean for the measured cycle, include frequency rates, percentage of safety
- Compute the standard deviation and multiply it by 3 to establish upper and lower confidence limits and identify the data above or below the upper confidence limit (UCL) or lower confidence limit (LCL). These are usually special cause variations
- If special cause exist, investigate them and develop correction action plans
- Develop upstream measures of common cause, and analyze the data

Statistics Definitions

Trends: Upward-downward change in reported data
Clusters: Indicate unusual changes in the system

Cycles: Indicate repetitive variation due to changes in process, for example, seasonal climate, weather

The current position of an organization with regard to QEH&S can be established by means of an initial audit/review of its management system and program elements.

- Identify gaps between any existing systems in place and the facilities guidelines
- Identify any significant hazards associated with the organization's operations
- Assess the level of knowledge and compliance with OHS standards and legislation
- Compare current procedures with best practice
- Review past experience with incidents and results of previous assessments and
- access efficiency and effectiveness of existing resources devoted to OHS management

Statistical Process Control

Traditional data basis may be too small because of under reporting of consequences. In incident reports, maintenance reports, rework, and environmental release reports. Three basic process questions should be asked.

- Was the defined process in accordance with the SOPs?
- What was the worker doing when he was injured or when the process failed?
- What should the process be like or what should the worker have done to prevent the injury or process failure? (perform a process risk assessment or job safety analysis)
- What primary process inputs (see Fig. 1) were inadequate or have failed in contributing to or causing the process nonconformance?

What defines critical process performance? Is there a process flow diagram illustrating the expected process for all critical jobs? This knowledge forms the foundation for SOP's, training, and observation inventories. Operation definitions must be clear, specific, and concise. This format is necessary for understanding training and measurement. You can then develop observation inventory sheets with columns for controls, resources, adequacy of existing procedures, proposals to rewrite SOPs or redesign the process or controls. (See Fig. 22.)

In reality, nothing is risk free until you are dead, and all uncertainty or risk for you is now zero, it's over! Risk will always range somewhere from zero to one, as a probability measurement. The challenge of quantifying risks is in determining with reasonable accuracy that historical data represent future loss events based on the frequency and severity of past events.

Measurement and statistical evaluation of data are important. Deming and Juran were statisticians and built the quality processes on a foundation of SPC.

You need a road map or process, and statistical evaluations define the forest from the trees and allow you to illuminate the path through the forest. Without SPC, management cannot distinguish between random process variability common cause and statistically significant deviations from the system or special cause, which generates a reactive, demoralizing and ineffective management system.

If managers are not careful making decisions based on statistical data, they will tend to react to slight shifts in the wind by over-steering and over-correcting, which in turn sets them up for an over-correction in the opposite direction. Deming emphasizes that 85% of the variability in process outcomes are due to common cause or those things built into the system. This same concept is more commonly recognized and accepted in the 80:20 principle in the Pareto principle, or principal of the vital or critical few.

'To be effective, you must decide on what you will measure and how to efficiently perform the measurement and data collection process. In addition, management must be prepared to execute a systematic process to determine the multiple root causes of significant process nonconformances or deviations. Two categories of data are used to measure organizational and process performance.

Qualitative — management information collected through audits, qualitative risk assessments, and inspection processes. Some of the data collected in the audit process can be scored by using a relative ranking system and thus become semi-quantitative. As mentioned in the early part of this book, Admiral Rickover maintained very good control over the Navy's nuclear power program by using these tools.

Quantitative — data collected through a process of measurement and direct observations to provide management with the information to avoid knee-jerk reactions to random fluctuations in the process. It is a tool to systematically engineer and control the continuous improvement process, prioritizing the allocation of limited resources to those areas that will control the root causes of process nonconformance and losses and have the greatest return on resource allocation–investment.

The probability estimate of risk for managers and underwriters is calculated by using statistics to estimate the future expectations of loss frequency and severity. To do that we must have a distribution of historical or representative data to create a loss probability distribution.

The range of potential losses around the expected average loss measures variability or the uncertainty of risk. The closer the values are distributed around the mean of the data distribution the greater our confidence that the data predicts future losses. By calculating variance and then its square root we determine the standard deviation, or confidence limits, as a range around the average.

The 95% confidence intervals are 2 standard deviations around the average, while the 99% confidence interval is 3 standard deviations around the mean. Because variance calculations square the data, they create a very large meaningless number. However, by determining the square root of the variance you can determine the average deviation between the mean and the data.

TABLE 16. Sample Loss Probability Distribution

Loss Value	Loss Frequency	Probability
$8.42 K	15	15/60 = 0.25
$24.9 K	10	10/60 = 0.17
$43.34 K	8	8/60 = 0.13
$11.44 K	7	7/60 = 0.12
$35.03 K	8	8/60 = 0.13
$9.78 K	12	12/60 = 0.2
Average Loss = $15.3 K	Total loss events = 60	

Average of expected future losses = Total of Losses/Number of Losses or Sum of the Loss Categories multiplied by the Category Probability
[($8.42 K)0.25] + [($24.9 K)0.17] + [($43.34 K)0.13] + [($11.44 K)0.12] + [($35.03).13]+ [($9.78 K)0.2] = [$2.1 + $4.2 + $5.6 = $1.4 + $2] = **$15.3 K**

As you can see these two sets of data in Tables 17a and 17b reflect the same average loss, but the second set has much less variability. Therefore, the standard deviation is much less. Which set of data reflects greater risk and less control? Which one would cost more to insure?

As mentioned earlier, people and the public in general have a difficult time understanding risk and putting it into perspective. In addition some companies have found that to be truly successful they must look at reducing processing variability and controlling nonconformances to within 6 standard deviations, which is 3.4 defects or nonconformance falling outside the control range for every million measurements.

To further illustrate a perspective of risk at 3 standard deviations or 99.9% conformance to specifications, the following losses would occur next year in the United States.

TABLE 17a. Sample: Data Set 1 (Same Avg. Loss as Data Set 2, Below)

Loss Value	Avg. Loss	Difference Loss Value – Avg. Loss	Difference Squared	Loss Frequency	Total Difference
$8.42 K	$15.3 K	–$6.88 K	47.3	15	$710 K
$24.9 K	$15.3 K	$9.66 K	92.2	10	$922 K
$43.34 K	$15.3 K	$28.04 K	786.2	8	$6,289.6 K
$11.44 K	$15.3 K	$3.86 K	14.9	7	$104.3 K
$35.03 K	$15.3 K	$19.73 K	389.3	8	$3,114.4 K
$9.78 K	$15.3 K	–$5.52 K	30.5	12	$366 K

Variance = $191.8 K
Standard Deviation = $13.85 K
96% Confidence Interval = $1.7 K – 29.2 K
99% Confidence Interval = 0 – $42.8 K
The range on the above data is quite large, reflecting the greater variability in the data or observed losses. This range also reflects a process that needs additional control. Data that have greater variability around the mean expected loses represent a less certain definition of risk losses.

TABLE 17b. Sample: Data Set 2 (Same Avg. Loss as Data Set 1)

Loss Value	Avg. Loss	Difference Loss Value – Avg. Loss	Difference Squared	Loss Frequency	Total Difference
$13.7 K	$15.3 K	–$1.6 K	2.56	15	$38.4 K
$15.8 K	$15.3 K	$500 K	.25	10	$2.5 K
$12.2 K	$15.3 K	–$3.1 K	9.61	8	$76.9 K
$19.9 K	$15.3 K	$4.6 K	21.16	7	$148.1 K
$21.1 K	$15.3 K	$5.7 K	32.49	8	$259.9 K
$12.5 K	$15.3 K	–$2.8 K	7.84	12	$94.1 K

Variance = $10.3 K
Standard Deviation = $3.21 K
96% Confidence Interval = $12.1 K – 18.5 K
99% Confidence Interval = $8.9 K – $21.7 K

- 12 newborns will be given to the wrong parent
- 18,322 pieces of mail will be mishandled per hour
- 2,000,000 documents will be lost by the IRS
- 315 entries in Webster's dictionary will be misspelled
- 20,000 incorrect drug prescriptions will be written this year
- 291 pacemaker operations will be performed incorrectly
- 3,056 copies of the *Wall Street Journal* will be missing a section.

—*(Mark Andrew, Liberty Risk Services, Quest 99 paper)*

Six Sigma, Upper Confidence Limit(UCL)/Lower Confidence Limit(LCL)

If you are not keeping score, you are just practicing.

—*Vince Lombardi*

Contrary to the historic use of 3 standard deviations above or below a fixed point, Six Sigma requires 6 standard deviations around the medium. Six Sigma was started by Motorla as a way to ensure a competitive advantage over the Japanese in the electronics business by focusing on product cycle time and quality. The *Financial Times* defined Six Sigma as an initiative aimed at the near elimination of defects from every person, product, and transaction (Hoerl, 1996).

The power in Six Sigma is that companies using it are controlling defects or nonconformances down to 3.4 per million opportunities. It has also been expanded for use in financial services in General Electric and has therefore been proven to have value in controlling a wide range of process risks beyond the production floor. Companies see the cost of quality as 20% of sales and therefore believe that to maintain their competitive market edge they must reduce their costs and/or losses by using management systems such as Six Sigma. This process is a means of looking at how a business achieves its goal or outcomes and offers strategies for process improvement.

To achieve this level of process control requires a heavy reliance on SPC and an equally strong commitment, including active participation in the process by executive management. General Electric's CEO, Jack Welch, believes in Six Sigma so strongly it is an integral component of the business planning process, and all managers who expect to be promoted to executive must be trained in Six Sigma, and have documented project success. This commitment is huge, requiring 4–6 weeks in the classroom over 12 months while working on approved projects.

Six Sigma requires bottom-line outcomes for projects and full-time leaders "Black Belts" (Masters) to lead the process. The process relies on statistical methods for determining the knowledge to achieve better facts and more profitable results and return on investment. It is designed to achieve breakthrough improvements by focusing on "out-of-the-box" thinking. Black Belts must accomplish at least four projects per year, delivering at least $500,000 to the bottom line. According to Reference 14, about 30 of America's largest companies have implemented this process. These companies include leaders in their businesses such as General Electric, Motorola, Avery Dennison, ABB, Lockheed Martin, Seagate Technologies, Raytheon, and Allied Signal.

Six Sigma is becoming a common language for different business units in the companies who have implemented the process. It forces them to share information and project outcomes and in general learn from one another. "It is amazing how little General Electric businesses knew about their customers prior to Six Sigma. The emphasis or metrics means customer needs and current performance in meeting them must be quantified and documented" (Hoerl, 1996). Using financial measures to select projects and measure success requires greater integration between the quality function and the rest of the business, including finance and procurement.

If we consider that waste is being generated anywhere work is accomplished, we can create a vehicle through which organizations can identify and reduce waste. The goal is total elimination of wastes through the process of defining it, identifying its source, planning for its elimination, and establishing permanent controls to prevent its reoccurrence (Breyfogle, 1999).

The seven elements to consider for the elimination of "muda" or waste are correction, over-production, processing conveyance, inventory, motion, and waiting. The initiatives to resolve waste can involve clean-up, storage or organization, standardization, training, and discipline (Breyfogle, 1999).

Six Sigma requires that work centers establish control plans to ensure processes are operated to meet or exceed customer requirements. A control plan is an extension of failure mode and effects analysis (FEMA) or other systems safety process evaluations to identify the multiple root causes of potential threats or risks and to define appropriate controls or process changes to eliminate or control the defined risk. The control plan provides a map or systematic way for identifying and controlling process risk. The system requires that we define the input requirements and necessary steps to meet output specifications and customer-required outcomes and the elimination of known or potential deviations.

SPC determines common cause (chronic) versus sporadic problems that oftentimes cause a knee-jerk reaction with management leading to additional over-reaction and correction, which increases rather than reduces process variability and costs. SPC charting is therefore useful for monitoring process stability, and it distinguishes between random events and common cause errors.

Juran's control sequence is basically a feedback loop with the following steps.

- Choose the control subject, what process you intend to improve
- Select units of measure
- Set control standards for the subject process
- Measure the process performance
- Interpret the differences between measured actual performance and the selected control standard
- Take action as necessary to move actual performance closer to the defined control standard (Breyfogle, 1999)

Common Cause or chronic problem breakthrough solutions or the sequence for breakthrough is described.

- Convince/enroll others in the belief that a breakthrough is needed
- Identify a few critical prioritized projects
- Organize or define the organization's systems that are necessary for the knowledge breakthrough
- Collect data to analyze the facts supporting required actions
- Evaluate the effects of proposed changes on the people and involve them to the extent possible to overcome resistance
- Take action to implement change
- Establish controls to ensure performance to the new standard

Problem solving and decision making require these factors.

- Problem recognition and definition, including multiple root causes
- Risk ranking, including financial consequences
- Evaluating control alternatives and their consequences, including return on investment and payback period
- Select best control alternative evaluating the level of expected risk mitigation and the cost–benefit analysis of the control alternatives
- Implement the plan
- Evaluate and provide feedback, measuring results

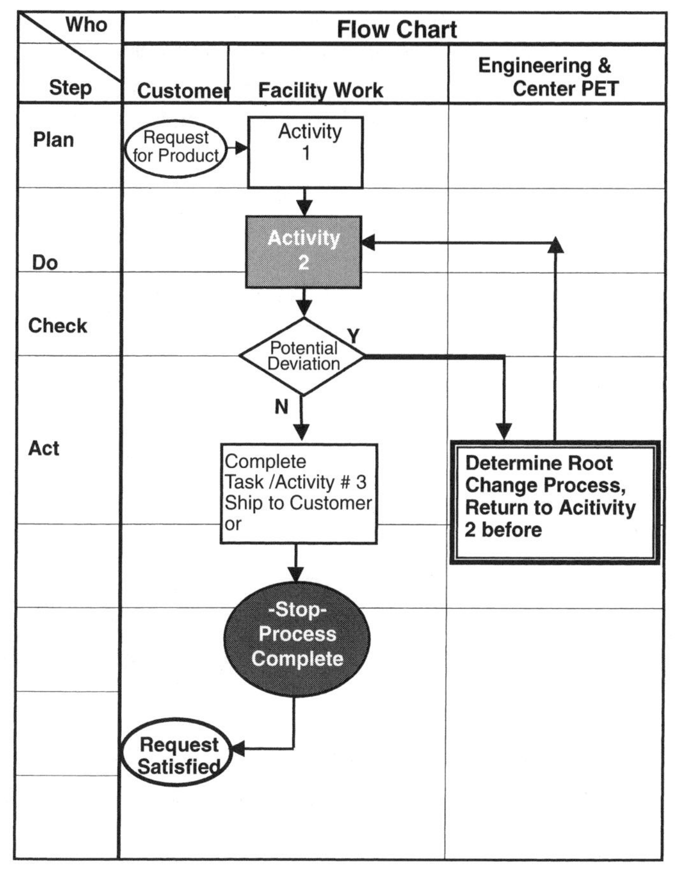

Figure 33 *Sample Process Flow Chart Model.*

In summary Six Sigma project plans look very much like the various risk identification and control strategies discussed earlier but with the advantage of having much better defined measurement data to assist with the decision-making process. Six Sigma plans generally include the following.

1. Identification of critical customer requirements, prioritized risks, and opportunities for improvement
2. Identification of stakeholders, select teams
3. Description of the business impact, loss frequency, and benefits in financial terms
4. Determine the metrics that will be used and begin process outcome measurement
5. Determine long-term process capability and performance using Pareto charts to help identify the locations or defects causing 80% of the problems, beginning to identify potential control alternatives

6. Define the process: flow chart and risk analysis and cause and effect analysis by using Ishikawa diagrams or other process flow models, FEMA, or MORT
7. Monitor the process, collecting and charting data t to identify special cause excursions in the process outside defined upper and lower control limits
8. Verify and documenting process improvement
9. Systematically plan (among teams) to improve processes that cause nonconformance, rework, delays, redundancy, breakdowns, and costly process common cause variation
10. Establish a data review and feedback system for continuous improvement

Process flow diagrams use the following symbols to define steps in the process (see Figure 33)

Terminal symbol or process stop-start points

Activity symbol describes a process step or task

Decision symbol asks a question following an activity, causing the flow chart to branch into two or more paths depending on the answers to the question or decision point.

Process Conformance Observation Sheet			
Observer: ____________			Date:__________
Location:____________			
Observation of: Critical Process Task Performance, use of tools, materials, & equipment			
Task - Tool - Material Description	**# of Conforming Tasks**	**# of Non-Conforming Task**	**Comments-Corrective Action:**
Task/Tool Requirement			
Task/Tool Requirement			
Task/Tool Requirement			
Task/Tool Requirement			
Task/Tool Requirement			
Task/Tool Requirement			
Task/Tool Requirement			
Task/Tool Requirement			
Task/Tool Requirement			
Sub Totals:			
Percent Conformance: Total Observations/Total Conforming & Total Non-Conforming Observations X 100 =			

Figure 34 *Task Observation Record.*

Corrective and Preventive Actions

Organizations should have formal documented reviews of process failures and incident nonconformances. These process failures should be analyzed to determine the multiple root causes (See Chapter 8). In addition, following-up audits or inspections to verify implementation of recommended corrective/preventive actions are essential for good performance and continuous improvement.

The responsibility for investigating incidents and process nonconformance must be established by policy and should include work center supervisors and not just technical or operations engineers. Management should conduct routine audits, including cross-facility reviews of the QEH&S procedures, equipment, or process inspections and can include critical job observations such as those performed with behavioral safety job observations.

Management system policy should specify the frequency of monitoring or measurement activities. It should also establish a process for identifying significant processes, systems, or facilities to evaluate, including the development and use of internal audit protocols. Audit and inspection responsibilities must be assigned to experienced personnel, with their competency documented through training-certification and/or experience. Management should also establish procedures to ensure that audit and inspection criteria are current with regards to regulations and internal policy or industry standards.

Operations supervisors, managers, and quality staff should establish documented procedures to inspect products or a service at appropriate stages of production or development to ensure that they meet design specifications. Those components of the service processes that do not conform must be held until they meet specifications. Finished products and service deliverables must be inspected to verify that they conform to customer requirements. These documented inspection results should include as appropriate SPC techniques and product verification, including critical job or task observations for conformance to QEH&S performance.

Executive management must review the management system at defined intervals to ensure its continuing suitability and effectiveness to satisfy the organization's QEH&S objectives. As appropriate, review outcomes should be incorporated in management action planning and be considered in the periodic management review process.

For the management system to be effective, it is essential that the organization's employees and other stakeholders "own" the system and its outcomes. The objectives and targets should be understood and supported by the organization's employees. Employees should be encouraged to accept the importance of their achievement in terms of both the organization's QEH&S performance and the benefits the process brings to the environment in which they work (profitability, etc.).

Figure 35 *The Non-Conformance Detective.*

Action Plan

- Establish periodic senior management review and evaluation process for the implemented QEH&S management system
- Establish audit, job–task analysis, and performance management system observations
- Evaluate SPC
- Evaluate Six Sigma quality management system for your most significant production and finance administrative procedure
- Determine benchmarking standards, both internal and external. and the performance review process for comparing with benchmark standards
- Develop a list of critical procedures, tools, and equipment
- Establish a maintenance schedule for critical tools

8

Nonconformance and Incident Investigation

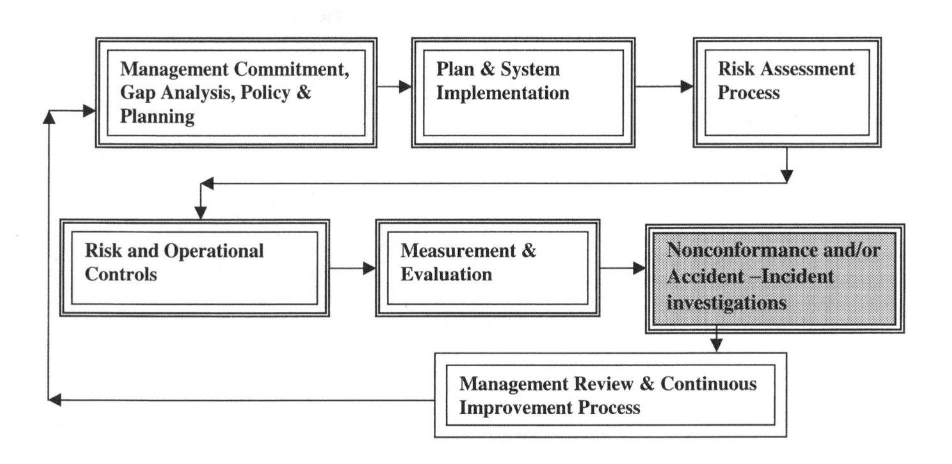

Our problems are man made; therefore they may be solved by man. No problem of human destiny is beyond human beings.

—*John F. Kennedy*

Problems are opportunities in work clothes.

—*Henry J. Kaiser*

You must establish a system to evaluate and determine the multiple root causes of customer complaints, incidents, and product or process nonconformances, and then develop changes in the process design to correct/prevent a reoccurrence.

In many ways it's like sailing; you set a course with your management system, but as the direction of the wind or the (organizational) current changes, it will be necessary to continue making changes in direction, tacking close to but not directly into the opposing wind.

Nonconformance and accident/incident investigations are a way to determine the multiple root causes in the system that promote substandard job performance. To be effective, the results of these investigations and the lessons learned must be shared with all potentially affected stakeholders, including employees and similar facilities or process work centers that could experience the same problem. As you proceed further along the course of continuous improvement the "tacks" or necessary changes will be fewer and less significant.

The intent of this chapter is to provide tools to help you understand the multiple root causes of process failures or accidents. If you can define a process and understand why it failed (Note: There are usually multiple causes in the system for the nonconformance), then you are in an excellent position to improve the faulty process and to share the lessons learned" with other stakeholders or organizational entities with similar risks. This element is key to the continuous improvement process. As I mentioned in Chapter 1, too often employees are blamed for process failures and accidents. I hope that after reading this chapter and trying multiple root cause analysis on some incidents or quality issues, you will start to see how the environment or substandard conditions and the multiple deficiencies in the system came together at the right time to cause the breakdown. This chapter provides an opportunity to learn from past mistakes.

Misunderstandings related to the term "Incident":

- "Every incident has a single cause."
- "Breaking rules and regulations causes incidents." (It's the employee's fault for getting killed)
- "Fault and cause are one and the same when explaining incidents."

Note: As a point of clarification, "incidents" has become more common term for accidents. However, historically, incidents have often been defined as a non-serious accident with no personal injury, whereas accidents were generally considered to include personnel injury or more serious damage to property. Incident in the context of process nonconformance is a more inclusive term.

PROCESS FAILURE/INCIDENT INVESTIGATION POLICY

Investigation of process failures must be done for these reasons.

- Process Nonconformance and accidents are caused.
- Process Nonconformance and accidents can be prevented if the causes are eliminated.

- Unless the causes are eliminated, process nonconformance and accidents will be repeated and ultimately become more serious, threatening the lives of employees and the viability of the business.

Most companies tend toward a culture of blame when it comes to determining the cause of incidents or process nonconformance. They don't understand or recognize that Joe's substandard performance or behavior was most likely caused by substandard conditions in the management system, including process design, communications, and leadership. Joe is therefore forced to deviate or use the wrong tools.

The primary reason why process failures are investigated is to prevent recurrence. Prompt and thorough investigation helps identify the multiple root causes of the incident and to establish the true costs of process nonconformance and accidents. It also creates effective change opportunities or improves process control. Process Nonconformance/Accidents investigations identify the risks or deviations in the system that need correcting. Corrective actions should also help reduce the risk of safety losses and personal injury, since safety problems often have the same basic root cause as quality and environmental health and safety (QEH&S) risks. I recommend that all process nonconformance and accidents be evaluated to determine the multiple root causes of process errors. Process failure investigations, which quickly assign errors to workers, fail to uncover system deficiencies. The attached Incident/Process Failure Investigation forms are used to investigate and determine the multiple root cause of incident/process failures. When the investigation has been completed and the root causes have been determined appropriate changes can be made to the work procedure or process to prevent a reoccurrence of the nonconformances.

The objective of these investigations is to determine how the system failed, and to implement the most appropriate or best corrective/preventive actions. Process failure/incident investigations are a methodical examination of an undesired event that did, or could have resulted in risk, physical harm to people, damage to property, rework or lose of product or service quality. Investigations are directed toward defining the causes of the process nonconformance and accidents and developing actions to control them.

WHY INVESTIGATE ACCIDENTS OR PROCESS NONCONFORMANCES, INCLUDING NEAR-MISSES OR INCIDENTS?

- Identify multiple root causes
- Develop corrective/preventive measures
- Audit effectiveness of QEH&S process/system
- Identify training needs or communications problems
- Measure supervisor's safety performance
- Show management commitment to continual improvement

CUSTOMER COMPLAINT/REWORK AND ACCIDENT, INCIDENT, AND ILLNESS DATA AS QEH&S PERFORMANCE INDICATORS

Process nonconformance or incident investigation data and information are vital, as they are a direct indicator of process performance and a flag for the conditions that lead to substandard performance or behavior. However, cautions exist relating to their use, therefore, proactive job observation data such as data collected from behavioral-based safety or process performance check sheets are more valid measurements of the system's conformance to process requirements.

- Most organizations have too few injury accidents or cases of work-related illness to distinguish real trends in nonconformance from random effects.
- A time delay will occur between QEH&S management system failures, harmful effects, for example, many occupational diseases have long latent periods. It is not desirable to wait for harm to occur or be recognized before judging whether the management system is working. In addition, studies have repeatedly documented that for every serious injury or process failure there are hundreds of near-misses, incidents, or minor injuries. Random circumstances and multiple causes contribute to the severity of accidents or process failures.
- If there is more per person output, increased workload alone may account for an increase in incident rates; in addition, many incidents (near-misses) are not even recognized.
- The length of absence from work that was attributed to injuries may be influenced by factors other than the severity of injury or occupational illness, such as poor morale, monotonous work, and poor management/employee relations.
- Accidents are often under-reported (and occasionally over-reported). Levels of reporting can change and can improve as a result of increased workforce awareness and better reporting and recording systems.

Process failure/incident investigation must have the following:

- Record of the assessment
- A means to identify loss, including direct and indirect costs
- A means to identify the multiple root causes of the risk or process failure
- A means to identify corrective/preventive actions to prevent further risk

Nonconformance documentation must include the following.

- Date, time, location
- Description and analysis of nonconformance and its multiple root causes
- Interim corrective/preventive actions
- Photos or diagrams

- Both direct and indirect costs, including assumptions used if costs are estimated
- Follow-up actions on all interim and permanent process changes or corrective/preventive actions

PROCESS FAILURE/INCIDENT ANALYSIS

Near-misses should be investigated. A comprehensive incident investigation program provides a broad base of information from which accurate conclusions can be drawn to help prevent process failure risk. Analyzing process failures and near-process nonconformances provides data for measurement, problem solving, feedback, and control. Process failures and near-misses (incidents) show that there are system problems that must be solved.

Management and the process enhancement team's PET basic goals are to evaluate and control process risk by identification of critical tasks; setting standards for work performance; measuring performance to those standards; evaluating the results of performance measurements, correcting substandard performance, including the substandard conditions causing the behavior; and commending–reinforcing desired performance. Process failure/incident analysis plays an important part in the control system.

Frequently managers blame the injured employee. They determine that the incident was caused by carelessness and that an appropriate corrective action is discipline and re-training. This approach is a quick way to get the incident investigation process completed, however, it falls short of identifying the multiple root causes and will guarantee that other nonconformances will occur and with the risk of more serious injuries or damage.

The management system that was the precursor to the substandard performance and the incidents totally ignored and missed the opportunity to correct it or prevent a reoccurrence. If I review incident or nonconformance reports that blame the careless worker and recommend training as the corrective/preventive action, I will automatically reject the investigation report. Reports that only identify a single root cause and blame the employee in most instances have been hasty and will not reflect an understanding of the multiple root causes and therefore will not be able to recommend effective corrective/preventive actions.

THE CARELESSNESS MYTH

How many people care less about being injured? Fraud or system abuse is always a possibility, but rarely carelessness. If you analyze most accidents, you'll find more management errors than worker errors and that fraud is rare and difficult to prove. If the multiple root causes of process failure are identified, you will see that the system failed and forced substandard performance, which resulted in an incident or loss of product, service, or rework.

Although it is true that the employee may have deviated from a recommended process, management has usually established conditions that forced the deviation. Some cases of workers' compensation are clearly abuse of the system. But this finding is a relatively small percentage of cases and in itself may be pointing to managerial issues that need attention. It is simply not true that employees care less about getting injured!

Many times process nonconformance or incident investigations stop after determining a single cause versus the multiple root causes for process failures. Unless managers understand the concept of multiple root causation and analysis, they will not be able to effectively evaluate and prevent the re-occurrence of process or systems failures.

Managers who hastily attribute the cause of process failure to carelessness miss the opportunity to uncover and correct/prevent significant process deficiencies and will not come up with effective solutions. Process failures and accidents can be eliminated if the causes are identified.

KEY ASSESSMENT CONSIDERATIONS

Identify multiple root causes, not symptoms (these causes are the conditions or process impacts that led to the substandard employee behavior or performance)

1. Purchasing
2. Methods (process instructions, etc.)
3. Tools and equipment
4. Environment
5. Personnel (including management and upervision)
6. Facilities
7. Materials
8. Maintenance
9. Training
10. Communications

Investigation Objective

Using the investigation tools provided at the end of this chapter determines how the management system failed and created the substandard conditions and the resulting employee behaviors and how to initiate corrective/preventive actions to prevent a reoccurrence.

Fact-Finding Process

- Conduct interviews
- Take samples

- Examine equipment
- Photograph and diagram the scene and process flow as it was intended and the points in the process that were deviated
- The process flow diagrams must demonstrate the original or intended process design and initial planning process and must point to the process deviated (and why) causing the incident or nonconformance.

Interview Techniques

- Conduct the interview as shortly after the nonconformance as possible
- Create a relaxed atmosphere, avoid blame, get all perspectives, and request ideas for prevention
- Keep the interview private to avoid group biases
- Look for facts and beware of smoke screens
- Ask open-ended non-leading questions
- Listen, test, and investigate all evidence
- Repeat the story back, probe into all aspects of the nonconformance or accident and get all sides of the story

Investigate

- The Physical Situation (What? Where?)
- The Person (Who?)
- The Time (When?)
- The Method (How?)

Equipment, Machines

- Were machines, equipment, and tools adjusted and properly guarded, maintained, and positioned correctly? Were the right tools for the job readily available and designed correctly?
- Have material handling and the ergonomic stress or risk and personnel process interface been considered?
- Are lighting, temperature, and vibration needs/risk considered, and are they appropriately controlled in the process?
- Are floor surfaces designed to minimize slips and falls?
- Has the process layout been designed to minimize physical stress and maximize efficiency?
- Are the generation of vapors, noise, or other occupational health hazards controlled?

The Person (Who?)

- Personnel placement and job qualifications procedures for the job have been established to ensure that the right person is hired for the right job

- Clear instructions and process communications have been established to minimize the risk of substandard performance
- Individual health, strength, endurance, flexibility to perform physically demanding work is considered when hiring or in the placement process
- Employee fatigue and physical stress as causes of ergonomic or stress claims or as an element to enhance process efficiency
- Considering unusual job conditions such as rush job, seasonal, or one-time-only when evaluating process nonconformance

The Time (When?)

- Who was involved?
- What happened?
- Where did it happen?
- The Method (How?) did it happen, and what were the multiple root causes of the process nonconformance?
- Did the employee use the correct procedure and were training, supervision, or other communications adequate to ensure process conformance?

RECOMMENDATIONS AND FOLLOW-UP

- Develop specific corrective/preventive actions
- Address all identified multiple root causes
- Identify those persons who are responsible for corrective/preventive actions
- Set corrective/preventive action target dates and follow-up process

PROCESS NONCONFORMANCE AND INCIDENT INVESTIGATION PRINCIPLES

Principle of Definition: A logical decision can be made only if the real problem is defined.

Principle of Multiple Root Causes: Problems are seldom if ever the result of a single cause.

Principle of Critical Few: In any given group of occurrences, a small number of causes will tend to produce the largest proportion of results (Parieto analysis)

Principle of Point of Control: The greatest potential for control exists where the action takes place. In other words, the first-line supervisor is the point of control for safety, quality, and production (Bird & Germain, 1996).

The primary reason for process nonconformance investigations is to prevent a reoccurrence. Prompt and thorough investigations help establish the multiple root causes of the nonconformance and their true costs. These investigations also create opportunities to control process failure losses. Analysis of accumulated nonconformance data also helps to identify patterns and multiple root causes of process breakdowns. Nonconformance or incident investigations identify the risks or process deviations that were taken and need correcting.

Supervisors and managers must be taught the difference between observing and seeing. First-line supervisors should inspect their job sites or work centers daily and document a formal inspection/observation at least weekly by using specific observations check sheets (see Chapter 7). Behavioral safety or performance management programs should have checklists of critical process requirements. Supervisors and employees should observe these jobs for process or system conformance. They should also evaluate corrective/preventive actions and plan conformance with the goal of creating an environment/system process that reinforces the desired behavior or performance.

Mid-level managers should accompany their managers on these inspections and use the same check sheets at least monthly, and senior management should accompany the inspection team at least quarterly or more frequently for new or high-risk operations.

Serious incident or nonconformance investigations should be conducted by a team. Supervisors may tend not to do a good job for several reasons.

- They will want to rush and blame employees rather than identify the systemic problems that may be caused or controlled by their boss.
- Multiple root cause analysis will identify problems or the need for corrective actions outside the supervisor's control. In fact, supervisors may identify problems owned by their manager and they may fear making their managers look bad if they document the true system problems with the failed process.

Quality improvement or safety committees can develop a list of the facility's critical tasks. Critical tasks are those that are an essential and routine element of the production process. They are also those tasks that as a group represent 80% of the facility's process risks.

It is necessary to teach the management team how to recognize these tasks and systematically analyze them by using the risk assessment process discussed in Chapter 5. It is also important for all supervisors and managers to understand how to perform job-task analysis and multiple root cause analysis to determine the causes of substandard performance. In most cases substandard performance is driven by substandard conditions beyond the employee's control.

Managers must recognize the incidents (near-misses) that ultimately lead to a serious process failure, including fatalities. The way to recognize incidents is by frequent job observations. Incidents are defined as the near-misses that do not result in personal injury, significant damage to the facility, or interruption of the production process. The ability of management to recognize and evaluate incidents will allow them to determine the weakness in the process or the potential root causes of a more significant process failure or injury. It is the "smoke" that if observed would allow you to carry out the changes before the "fire" gets out of control!

In general, most process failure analysis stops after a single and often simple cause is identified. Usually multiple root causes of process failures exist. Unless managers understand the concepts of multiple root cause analysis, they will not be able to effectively evaluate accidents, which in turn eliminates the possibility of preventing reoccurrences. Review Figure 18, which clearly documents the hundreds of warnings that are typically not investigated until a serious injury or process nonconformance occurs.

Process Nonconformance and Incident investigations are necessary because (Bird & Germain, 1996).

- They are caused.
- They can be prevented if the cause is eliminated. Unless the cause is eliminated, it will happen again and will ultimately have much more serious consequences.
- Accidents or near-misses demonstrate that there are process problems, which must be evaluated and controlled not unlike any other management or process risk.
- Process failures and accidents can be prevented if the causes and substandard conditions are eliminated.
- Unless the multiple root causes are eliminated, the incident/accident will repeat.
- Other accidents and process failures can be more serious, with possible fatalities or loss of business.

LEARNING FROM AND COMMUNICATING INVESTIGATION RESULTS

The organization, having learned from the investigation, should share the lessons learned from the investigation. Investigations should determine the multiple root causes of the following.

- Process nonconformance/accidents and their multiple root causes — these investigations will identify a need to communicate findings to management and employees of the organization and to share the documentation and planned changes with employees or facilities with similar process risks.
- Include identification/documentation of relevant findings and recommendations from investigations for considerations in the continuing improvement process.
- Implementation of interim and long-term corrective/preventive remedial controls should be followed up to ensure that timely and effective changes are implemented.

Nonconformance/Incident Investigations. All supervisors and managers should be trained in investigations, including multiple root cause analysis.

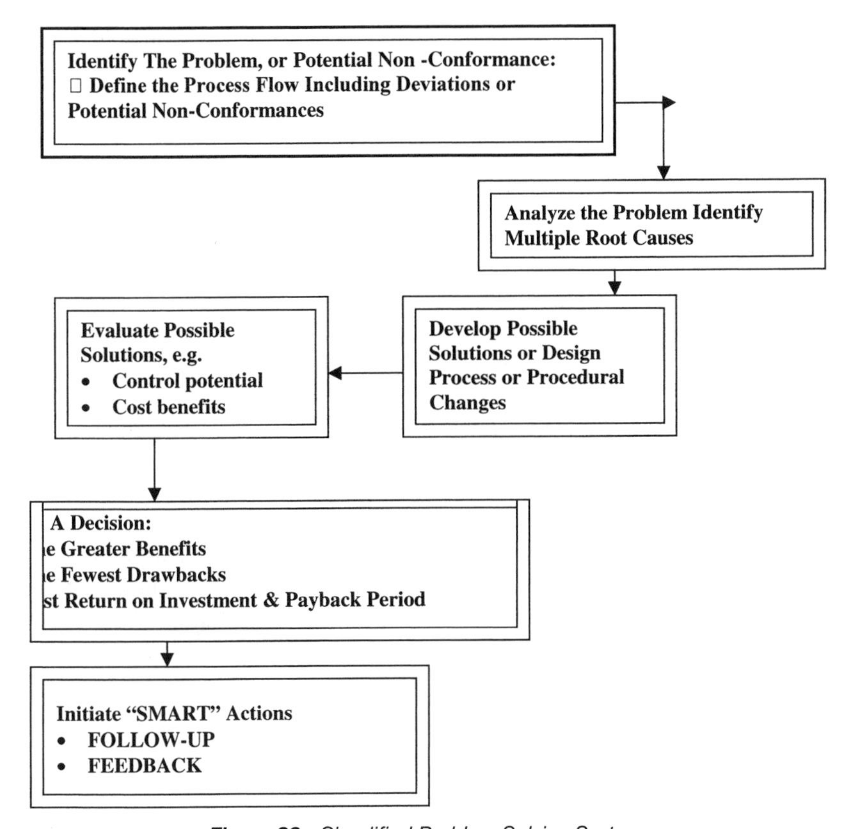

Figure 36 *Simplified Problem Solving System.*

Investigation reports must contain recommendations and a timetable for the implementation of corrective/preventive actions and should assign responsibility to designated personnel and managers. In most circumstances, investigations should include the responsible supervisor. In addition, corrective/preventive actions must be discussed with affected stakeholders prior to implementation, and the effectiveness of corrective actions should be monitored.

Some of the tools provided in this chapter, including the check sheets at the end, are intended to help determine the multiple root causes of process failure. In addition, investigators should be taught how to draw process flow diagrams or cause-and-effect diagrams. These diagrams will help clarify and define the process as it was executed and as it should be by clearly illustrating changes that may have caused the process failure or the need to change the process to prevent a reoccurrence. Process flow diagrams and change analysis are excellent tools to help identify the systemic causes of the process nonconformance.

It is critical that managers and employees working on process enhancement for QEH&S visualize the process and all the things that happen in it and in support of it (the myriad of sub-systems). Unless the system or process is well-defined and understood as an optimal designed system that can be compared with the existing operating process, the hope of clearly identifying and eliminating process risk is significantly diminished.

NONCONFORMANCE ACCIDENT–ILLNESS INVESTIGATION PROCESS POLICY

The organization should establish policy and procedures to ensure that work-related fatalities, injuries, illnesses and process nonconformances or environmental releases are systematically investigated. Emphasis should be placed on methods to identify the multiple root causes of the incident and to prevent their future occurrence.

Organizations should have procedures for reporting and documenting nonconformance or incident investigations. The prime purpose of the investigation procedure is to prevent future incidents. The occurrence of accidents and incidents, environmental nonconformances, rework, or other quality issues are evidence of a management system failure. Therefore, to find out why an incident or nonconformance occurred is the first step in implementing corrective/preventive actions in the continual improvement process. In addition, both direct and indirect costs of accidents or process failures should be documented, and lessons learned should be shared with all potentially affected parties.

PROGRAM SYNERGISM/CONCLUSION

The root causes of production quality control problems are often the same causes of EH&S problems. Process safety improvements will enhance the production process. Process efficiency or quality improvements implemented without a consideration for safety or environmental risks may increase workplace hazards, quality, and process inefficiency.

Like TQM or re-engineering, workplace loss control or safety requires that supervisors and employees understand work as a process. Managers and supervisors must be able to use tools from quality improvement processes to identify critical jobs or tasks. Job-task analysis should then include diagrams to help define the process steps, including resources applied to the process. Management can evaluate the process by using systems safety tools to identify the multiple root causes or systemic reasons for nonconformance. They must then measure performance with simple frequency charts and/or tracking systems, including the costs of losses per department, process, or unit of production.

Companies with integrated QEH&S systems will be in a much stronger position to focus on and reduce the costs and liabilities associated with workplace hazards. Incidents or near-misses identify process problems that must be controlled. Accidents or incidents are not unlike any other management or process problem; they require the same quality improvement approach and management commitment/leadership.

ACTION PLAN

- Train all supervisors and managers in multiple root cause nonconformance investigation methods
- Document direct and indirect costs associated with a nonconformance and charge that cost back to the work center with the failure
- Establish a system to share lessons learned with regards to the causes of process failures and the controls or process changes being made to correct/prevent a reoccurrence

Figure 37 *Which Way do We Go.*

TABLE 18a. Potential Process Nonconformance or Incident Investigation Questions

Who

Who was involved in the process failure?
Who regarded the process as working?

What

What was the process failure?
What was the loss (estimate or document direct and indirect costs)?
What were the employees doing and what where they told to do?
What tools and/or equipment were being used?
What were the required process and known process changes?
What verbal or written instructions had been given to employees?
What specific precautions were required and given?
What protective equipment was used or should have been used?
What had other persons done that contributed to the process failure?
What did the employee or witnesses do when the process failure occurred?
What did the witnesses see?

TABLE 18a. (*continued*)

What will be done to prevent recurrence?
What process requirements were violated?
What changes to procedures or processes are necessary?

When

When did the process failure occur?
When did the job start?
When was the risk or process failure discovered?
When did management last check the job progress?
When did the employee first sense something was wrong?

Why

Why did the procedure or process fail (identify multiple root causes)?
Why weren't protective equipment or barriers used, or lock-out tag-out procedures followed?
Why were the tools or machines contributing to the process failure used?
Why wasn't the manager/supervisor there at the time at the time of the nonconformance?

Where

Where did the process failure occur?
Where was the manager/supervisor at the time?
Where were other workers at the time?
Where were witnesses when process failure occurred?

How

How might the nonconformance, incident, or environmental release been avoided?

TABLE 18b. Process Failure Analysis. Personnel Selection, Training and Mentoring/OJT, Supervision: Did a Person's Lack of Physical or Mental Aptitude, Skill, or Knowledge Contribute to the Nonconformance?

Yes	No	Potential Nonconformance Causes:
		Was the person given a pre-placement medical examination?
		Was the examination based on a job safety and health and physical capability?
		Can the task or process be simplified?
		Did a pre-placement interview and orientation include a review of the work conditions and requirements, such as the handling hazardous materials or working in high places or confined spaces?
		Was the person given a general and work-site/process-specific safety orientation?
		Has the individual received formal job training, including EH&S?
		Was on-the-job training (OJT) provided? Was it renewed in the past year with any kind of updated process control training, briefings, or written material?
		If the person lacked physical or mental capability, were workplace modifications considered?
		Was the person given adequate supervised practice to develop necessary skills to perform the task?

TABLE 18b. *(continued)*

Yes	No	Potential Nonconformance Causes:
		Was the task performed frequently enough to maintain skill levels and familiarization with procedures?
		If any of the questions were answered *no*, this category may be one of the root causes of the incident/accident or process failure. Analyze the following.
		Do the hiring, placement, and training programs include adequate examination and evaluation of each person's basic capability to perform the job safely? Do they include provisions to modify the workplace or the activity to accommodate personnel with permanent or temporary disabilities?
		Are there adequate supervised demonstrations and practices to develop skills and procedural knowledge?
		Is there additional loss potential due to inadequate standards or compliance?
		Do the employee training, personal communications, and group meetings include adequate training to develop and update each person's knowledge of how to perform in conformance with QEH&S requirements?
		If personnel training standards are adequate, why weren't they complied? What additional loss potential exists due to inadequate personnel selection, training, and supervisor standards?

TABLE 18c. Personal or Job Stress: Was Physical or Mental Stress a Factor in the Incident?

Yes	No	Potential Nonconformance Causes:
		Was the person suffering any temporary illness that created personal stress or decreased abilities?
		Was the person using any medications or under the influence of any drugs, including alcohol?
		Were there any recent significant changes in the job (changes in tools, equipment, materials, pace, environment, etc.)?
		Was the task load, pace, or duration high enough to physically or mentally fatigue the person?
		Was the person exposed to environmental stress from: — vibration — noise — heat or cold — chemicals or contaminants
		Were movements constrained or were abnormal body positions required?
		Were any personal or family situations preoccupying the person?
		Were difficult judgments required while doing the task?
		Were there activities or conditions that might have caused distractions?

TABLE 18c. (*continued*)

Yes	No	Potential Nonconformance Causes:
		If any of the questions were answered *no*, this category may be one of the root causes of the incident/accident or process failure. Analyze the following. a) Are environmental controls, engineering controls, and job observation practices adequate to identify and control the effects of stress-producing situations? b) Are standards adequate to define the control measures and the level of performance needed to identify exposures and control the risks? c) If the standards are adequate, why weren't they complied?

TABLE 18d. Engineering: Was the Design of the Facility, Process, or Equipment and Tools Adequate, or was Inadequate Incorporation of Control Devices During Design or Fabrication a Factor?

Yes	No	Potential Nonconformance Causes:
		Were hazard assessments made during the conceptual design stage of the facilities?
		Were the hazards revealed by this incident or were accidents identified in the preliminary risk assessment?
		Were the effects of operations that interface with the process considered in the hazard analysis or risk assessment?
		Were all appropriate standards, codes, and past accident reports researched to establish criteria for design of the facility?
		Were controls for the risks incorporated into the facility design?
		Did the EH&S staff participate?
		Was a systems safety review performed with design and production engineering?
		Were the job or process requirements specified to the contractor or vendor?
		Were pre-construction conferences held with contractors, and did agendas for these conferences include QEH&S specifications during construction?
		Was the facility equipment examined to ensure that QEH&S criteria had been followed before acceptance of the facility?
		Were the initial operations of the system supervised closely for overlooked hazards or process risks?
		Were any modifications made to the original design, and were those modifications evaluated for risk during the modification design?
		If any of the questions were answered *no*, this category may be one of the root causes of the incident/accident or process failure. Analyze the following.

TABLE 18d. (*continued*)

Yes	No	Potential Nonconformance Causes:
		Are the purchasing and engineering controls adequate to ensure that potential hazards found are evaluated?
		Are environmental safety and health standards researched? Is past experience evaluated and are appropriate criteria incorporated into the design and construction of facilities and equipment?
		What is the additional loss potential due to inadequate process, standards, or compliance?

TABLE 18e. Materials and Procurement: Was Inadequate Purchasing, Management of Materials or Disposal of Scrap, Waste, or Obsolete Items a Factor?

Yes	No	Potential Nonconformance Causes:
		Were environmental QEH&S standards researched and specified to vendors of tools and equipment, as well as to contractors that built the facilities involved in the incident?
		Were the requirements for the materials used, as related to the accident, researched and were material safety data sheets obtained before purchase?
		Were the materials selected the safest ones that could be obtained?
		Was the coordination of requests for tools, equipment, or material purchase adequate?
		Were QEH&S information and instructions on equipment of materials communicated to the supervisor of the area?
		Were the hazardous material labels on material containers and safety placards or tools kept intact and legible?
		Were appropriate handling methods and personal protective equipment used when unloading, moving, storing, and issuing the materials?
		To prevent deterioration or instability, were materials and equipment stored properly prior to use?
		Were the hazardous properties of waste, scrap, and unserviceable or obsolete equipment identified so proper disposal methods could be followed?
		Are materials or equipment included in the QEH&S program, and are they adequate to ensure that potential hazards are identified before items are purchased or facilities are constructed and that risks are controlled throughout the life of facilities, equipment, and materials?
		If these materials and equipment acquisition activities are adequate, why weren't they complied?
		What is the additional loss potential due to the inadequate materials procurement compliance?

TABLE 18f. Maintenance (Facility, Equipment, and Tools): Was Premature Failure or Malfunction of Equipment or Facilities a Factor?

Yes	No	Potential Nonconformance Causes:
		Have service manuals, planned use, and government codes been checked to assess the needs for preventive maintenance of critical plant and equipment?
		Was a maintenance schedule written?
		Are adequate time and priority given to preventive maintenance?
		Was all preventive maintenance conducted according to the schedule?
		Were written maintenance instructions or adequate verbal instructions given?
		Was preventive maintenance work regularly audited?
		Are lubrication and service points readily accessible?
		Are the specified lubricants or approved substitutes used?
		Were parts assembled correctly?
		Were adjustable parts set to correct specifications?
		Are assembly and adjustment of parts made following clearly written standards in technical manuals or procedures?
		Are appropriate parts cleaned and resurfaced as required?
		Were the defects or deficiencies adequately described in a written repair work order?
		Was the mechanic given adequate written or verbal instruction of the repair(s) made?
		Were adequate time and assistance scheduled for the repair work?
		Was the equipment checked for proper operation when the repair was completed?
		Did a skilled craftsman or mechanical inspector check the repair?
		Were specified parts or approved substitutes used?
		Do the engineering controls, purchasing controls, and inspections include adequate research of requirements for maintenance?
		Do the job analysis, procedures, and personal communication include adequate maintenance instructions?
		Are the standards adequate to define the control measures and the level of performance needed to identify loss exposures and to control the risks?

TABLE 18g. Procedures/Standards: Were Inadequate Methods, Procedures, Practices, or Rules a Factor?

Yes	No	Potential Nonconformance Causes:
		Were QEH&S methods developed for the process, and were these methods communicated to all responsible managers, supervisors, and employee trainers?
		Was the procedure correct and current?

TABLE 18g. (*continued*)

Yes	No	Potential Nonconformance Causes:
		Has the procedure been reviewed and updated within the past 12 months?
		Did the procedure account for the results of a risk assessment or a task analysis?
		Were QEH&S work practices for the general type of work that was involved in previous incidents (lessons learned) published and available to supervisors and employees?
		If there were no procedures, were there QEH&S requirements to identify or control the risk?
		Were the requirements published and distributed to affected employees?
		Were the requirements reinforced with a sign or notice in the work area?
		Was the method, procedure, practice, or rule consistent with other instructions?
		Do the engineering controls, job analysis-procedures, and organizational rules include adequate study of work processes and methods development? Are risk assessments performed for critical tasks?
		Are the standards adequate to ensure that methods, procedures, or practices used are appropriate for the hazard and the evaluated degree of risk?
		If the program and standards are adequate, why weren't they complied?
		What is the additional loss potential due to inadequate procedures or compliance?

TABLE 18h. Management and Supervision: Was Inadequate Leadership a Factor?

Yes	No	Potential Nonconformance Causes:
		Was safety policy written and well-known to managers and employees in the area involved?
		Was control of the loss exposure affected by limited budgeting of personnel, equipment, materials, or other financial support?
		Was responsibility for control of the loss exposure related to the accident assigned to a specific person or position, and was sufficient authority given to enable effective control?
		Did the responsible person have management training in QEH&S related to the type of loss exposure?
		Were the potential risk exposure and its control measures discussed at a management meeting within the past 12 months?
		Does the QEH&S program reference manual provide adequate guidance that would control this type of accident or process nonconformance?
		Were program control measures that could prevent the loss exposure audited or evaluated within in the past 12 months?

TABLE 18h. (*continued*)

Yes	No	Potential Nonconformance Causes:
		Are the management training elements of the program adequate to ensure that planning, organizing, leading, and controlling QEH&S are effectively incorporated into the organization management program?
		If the program and standards are adequate, why weren't they complied?
		What additional loss potential exists due to inadequate program, standards, or compliance?

TABLE 18i. Tools and Equipment: Were Inadequate or Unsuitable Tools, Machines, or Vehicles a Factor?

Yes	No	Potential Nonconformance Causes:
		Were the tools used in activities that were related to the accident correct for the task, as specified in instruction manuals and QEH&S standards?
		Were tools used correctly, as anticipated when purchased or built?
		Were the tools or equipment in good condition and not worn beyond limits for safe and proper use?
		Were the equipment, tools, or facility used within an expected service life?
		Were adequate numbers of correct tools and parts in proper condition and available for the task?
		Was the equipment overloaded or operated at excessive speed, temperature, or any other abnormal way?
		Were tool or storage areas convenient and well-organized so the proper tools could be readily obtained?
		Was the contractor, or others who are inadequately trained, allowed to use the tools, equipment, and facilities?
		Were inspection frequencies increased as the tools, equipment, or facilities became older or their use rate increased?
		Was there a procedure for prompt repair, adjustment, sharpening, and other maintenance of tools and equipment to help keep them in safe condition?
		Have defective tools involved in the accident or process nonconformance been replaced and salvaged, made inaccessible, or made to prevent unintentional use until refurbished?
		If defective equipment was involved, has it been shut down, locked out, and tagged to prevent its use until its repair or replacement?
		Had the equipment defects been reported to supervision?
		If these activities are included in the program, are the standards adequate to ensure that safe tools and equipment are used?
		If the program and standards are adequate, why weren't they complied?
		What is the additional loss potential due to the inadequate program, standards?

TABLE 18j. Multiple Root Cause Analysis Risk Assessment Form (see Figure 18j)

TASK:

Plant Location: **Date:** **Risk Assessor/Team Facilitator** **Department:**

Step in Operation Task or Event	Potential Incident/ Accident; Unwanted Energy	Caused By:	Step or Task Changed	Probability	Consequence	Risk Rank	Current Controls & (Used, Not Used, Failed, Impractical)	Recommended Barrier Controls & Mgt. System Changes

Basic Risk Factors: Design, Communications, Supervision Training, Procedures, Tools & Equipment, Materials, Maintenance, Incompatible Goals

Risk Ranking: High Risk = (1) A risk which could result in serious injury or death, major health risk, or where several could receive lesser injuries, a serious fire threat, significant environmental liability/release, significant threat to market or financial loss. **Moderate Risk = (2)** A risk that could cause a serious incident/accident, loss of client, or significant damage to product or work in progress. **Low Risk = (3)** A minor injury or delay in the work process (Keep in mind Probability, see attached Risk Matrix)

Complete and return within 24 hours of incident.

<table>
<tr><td colspan="2">Name:</td><td>Employee #:</td><td>Date/Time of Incident:</td><td>Office Phone:</td></tr>
<tr><td colspan="4">Home Address:</td><td rowspan="2">Home Phone:</td></tr>
<tr><td colspan="2">Street</td><td colspan="2">City, state, zip</td></tr>
<tr><td>Dept. Name/Number:</td><td colspan="2">Job Title:</td><td colspan="2">Hire Date:</td></tr>
<tr><td colspan="3">Name Witness(es):</td><td colspan="2">Office Phone(s):</td></tr>
<tr><td colspan="5">Describe injury or illness.</td></tr>
<tr><td colspan="5">Describe in detail how the injury or illness occurred. What was being done? Where did the incident occur?</td></tr>
<tr><td colspan="5">Name any machine, tool, substance or object involved.</td></tr>
<tr><td colspan="5">Part of body affected (indicate right or left, front or back, etc.):</td></tr>
<tr><td colspan="3">Was first aid given? ☐Yes ☐No
If yes, date first absent: ____/____/____</td><td colspan="2">What other medical treatment, if any?</td></tr>
<tr><td colspan="3">Who gave first aid/other medical treatments?</td><td colspan="2">Where did treatment occur?</td></tr>
<tr><td colspan="4">How many days away from work? Explain why time off was required.</td><td>Date returned to work:
____/____/____</td></tr>
<tr><td colspan="5">What can be done to prevent this injury or illness from occurring again?</td></tr>
<tr><td colspan="4">Employee Signature:</td><td>Date:
____/____/____</td></tr>
</table>

Figure 18j *Employee Injury/Illness Report.*

TABLE 18k. Change Analysis Evaluating Barriers & Basic Process Risk Factors (see Figure 18k)

TASK:

Process: To more effectively identify the Basic Risk Factors & Multiple Root Causes of Process Nonconformance draft a simple process flow diagram depicting how the process was expected to happen and how it was actually done. For each step in the process that was deviated from or changed use the form below to accurately determine why it occurred and what needs to done to prevent a reoccurrence and to "Share Lessons Learned"

Plant Location: **Date:** **Risk Assesso/Team Facilitator** **Department:**

Work–Process Basic Risk Factors	Accident Free Process	Non-conforming or Accident Process	Difference and Why	Impact-Contribution to Accident or Potential Contribution to Future Process Failures
Energy Barriers				
Design				
Communications				
Supervision				
Training				
Procedures				
Tools & Equipment				
Materials				
Maintenance				
Incompatible Goals				
Process Flow				

COMPLETE ALL SECTIONS. Completed Form Must Be Submitted Within 24 Hours of Incident. Use and Attach Additional Paper, if Needed.

Please check one to indicate the appropriate category: ☐ **Accident (injury, illness, property damage)** ☐ **Incident (sequence of events that could have resulted in an accident)** ☐ **"Near-Miss" Incident (Incident which could have caused injury, illness or damage but did not)**

Print Employee Name	Employee Number	Date of Birth / /	Sex ☐ Male ☐ Female	Date Reported to Supervisor / /	Date Report Prepared / /
Department Name	Department Number	Date of Hire / /	Length on Present Job	Specific Location of Accident	

Day of Occurrence	Date of Occurrence	Time of Occurrence	List Name(s) & Extension(s) of Witness(es)	Treated At:
☐ *Monday* ☐ *Thursday* ☐ Sunday ☐ *Tuesday* ☐ *Friday* ☐ *Wednesday* ☐ *Saturday*	/ /	________ ☐ a.m. ________ ☐ p.m.		☐ *Facility Medical Department* ☐ *Physician* - Dr________ ☐ *Hospital* -________ ☐ *Specify Other* -________

Describe, in detail, how this event occurred:

Check All That Apply In Each Category

Injury/Illness Type		Affected Body Part		Immediate Cause(s): Unsafe Conditions	Root Causes: "Management System Latent Defects"
☐ Abrasion	☐ Hearing	☐ Head	☐ Internal	☐ Guarding	☐ Communications: Procedures/Training
☐ Amputation	☐ Hernia	☐ Ear	☐ Chest/Shoulder	☐ Horseplay	☐ Supervision
☐ Burn	☐ Radiation	☐ Eyes	☐ Abdomen	☐ Lifting	☐ Incompatible Goals
☐ Contusion (Bruise)	☐ Shock (Electrical)	☐ Mouth	☐ Leg	☐ Lighting	☐ Facility design, use,
☐ Crushing Injury	☐ Sprain/Strain	☐ Nose	☐ Foot	☐ Traffic.	☐ Tools/Equipment Use/Selection
☐ Cumulative	☐ Visual	☐ Neck	☐ Toe	☐ Ventilation	☐ Maintenance
☐ Trauma	☐ Multiple (Specify)	☐ Arm	☐ Multiple (Specify)	☐ Work Area Congestion	☐ Housekeeping
☐ Cut, Puncture	☐ Other (Specify)	☐ Wrist	☐ Other (Specify)	☐ Use of Personal Protective Equipment	☐ Work or Process Flow Design
☐ Dermatitis		☐ Hand		☐ Stable Body Use/Positioion	☐ Work Station Setup
☐ Emotional (Specify)		☐ Finger		☐ Unsafe nonconforming behavior	☐ Unauthorized Activities
☐ Fracture		☐ Hand		☐ Other (Specify)	☐ Pre-Job Assessment
	________ Specify		________ Specify		☐ Other (Specify) ________

IMMEDIATE CAUSE(S)	MULTIPLE ROOT CAUSE(S)
1A. What conditions of tools, machines or the work area caused this incident?	1B. Why did these conditions exist? What management system defects create the substandard conditions causing the risk?
2A. What did the person(s) involved do or fail to do?	2B. Why did the person involved do, or fail to do, something that caused or contributed to this accident?

Supervisor/Manager Corrective Action *(Check all that apply. If action pending, document on additional paper and attach.)*

☐ *Develop/Revise Job Practices*	☐ *Improve Orientation Training*	☐ *Install/Replace/Adjust Guards*	☐ *Improve Communications*	☐ *Other (Specify)* ________
☐ *Enforce Disciplinary Action*	☐ *Improve Housekeeping*	☐ Institute *Hazard Analysis*	☐ *Review via Task Force*	
☐ Improve *Emerg./Med. System*	☐ *Improve Maintenance*	☐ *Modify/Replace Tools/Equip.*	☐ *Revise Process/Equip./Layout*	

Specific Corrective Action(s)		Person(s) Responsible	Action Deadline
Print Supervisor Name	**Print Manager Name**	**To Be Completed By Safety Department** ☐ *First Aid Only* ☐ *No Medical Leave* ☐ *Modified Work Available* ☐ *Treatment Beyond First Aid* ☐ *Medical Leave* ☐ *Modified Work Unavailable* ☐ *Lost Time (+1 Full Day)* ☐ *Modified/Restricted Work*	
Supervisor Signature/Date	Manager Signature/Date		

Figure 18k *Supervisor's Incident, Accident and "Near-Miss" Incident Investigation Report.*

Supervisor's Incident-Injury Investigation Summary of Findings: What was the employee doing, and what was different about this instance over previous time the process was performed. What changed? Were barriers used? Provide a simple process flowchart and identify deviations or changes, and explain why they happened.

Recommended Corrective Action Plan & Documentation: Describe both the immediate recommended corrective actions to establish barriers between employees and the hazard, and process changes to improve the management of this process (eg. prevent "Management System Latent Defects"). Include target dates for completion of each action, and document completion on this form or with attached documentation.

Supervisor Signature & Report Date - Supervisor Initials & Action Plan Completion Date

EHS Reviewer Signature & Date

Figure 18k *(continued)*

Approving Manager's Signature & Date

Cost Estimates:

Damage Estimate	**Injury Costs:** Lost Time: Medical: Other:	**Damage to Tools/Equipment/ Facility or Product:**	**Supervisors Lost Time:**	**Total Est. Direct Cost:**	**Total Est. Indirect Cost:**
Corrective Action Cost	**Briefly Identify or List Specific Corrective Action Cost & The Total Costs For All Changes:**				

Figure 18k *(continued)*

9

Management Review and Continuous Improvement

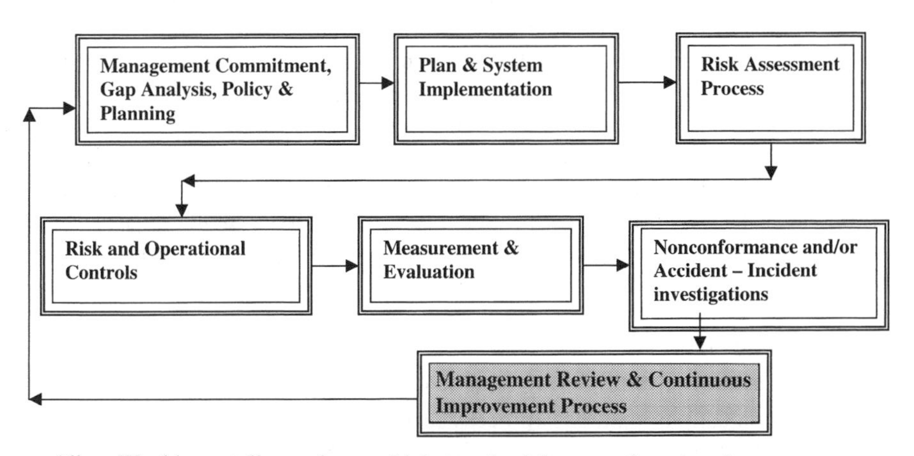

Alice: Would you tell me please which way I might to go from here?

Cheshire Cat: That depends a good deal on where you want to get.

Alice: I don't much care where.

Cheshire Cat: Then it doesn't matter which way you go.

—*Lewis Carroll, Through the Looking Glass*

Senior management must periodically review the quality and EH&S (QEH&S) management system to ensure that the process for continuous improvement is maintained, is effective, and meets both operational and regulatory objectives. Other management system progress reports, including cross-facility audits, consultant or registrar audit reports and recommendations, changes in regulations, and ISO or industry standards should be included in the integrated assessment review. The management review should document considerations regarding process changes, changes to policy and management system objectives, and milestones. In addition, the results of risk analysis, nonconformance investigations, systems, and product or production design reviews, critical process observation reports, statistical process control, quality warranty, or rework data will be valuable for the management review process.

This chapter reviews the various tools and sources of information and data that can be used to help senior managers facilitate the documentation and sharing of experience about the management system. The use of some or all of the measurement and evaluation tools discussed in Chapter 7 are critical aspects of your management system. Remember two details.

1. What gets measured, gets done.
2. Garbage in, garbage out.

The management review process is like having charts and a GPS system on your sailboat. You can use them and avoid sailing onto the rocks, also you can maximize the efficiency of your journey or go on blind luck. The choice is yours. With minimal effort you and your organization can access the data to do a better job piloting your organizational ship through rough seas. Without measurement and periodic management review, you will be like Alice with no idea of which direction to go. Continuous improvement requires data so you will know where you are, where you have been, and — most important — where you are going!

Integrated industrial work process QEH&S systems will ensure overall corporate success and maximize profitability and corporate viability. These steps in the continuous improvement process require management commitment and periodic review.

- Risk assessments
- Identification of multiple root causes of process nonconformance
- Communications methods (process instructions, standard operating procedures [SOPs], training, and supervision)
- Facilities, tools and equipment design, use, and maintenance
- Environment (including corporate culture and physical resources and processes)
- Material specification and purchasing process, storage, use, transportation, and disposal

CORRECTIVE AND PREVENTIVE ACTION

The findings, conclusions, and recommendations reached as a result of monitoring, auditing and reviewing the management system should be documented, and the necessary corrective and preventive actions should be identified.

Management should ensure that these corrective and preventive actions have been implemented and that there is systematic follow-up to ensure their effectiveness. In the event of a process failure, incident or nonconformance management must determine the multiple root causes of the failure. Appropriate corrective actions must then be taken and the lessons learned should be shared with affected employees or management.

The organization's executive management should, at appropriate intervals, conduct a review of the management system to ensure its continuing suitability and effectiveness in satisfying the organization's policy and objectives. The review of the management systems should be broad enough in scope to address the QEH&S implications of all activities, products, or services of the organization, including the following.

a. An evaluation of the suitability of policies and procedures
b. Review of objectives, targets, and performance towards meeting those objectives
c. Findings of the management systems audits, job–task observations, and inspections
d. An evaluation of the effectiveness of the management systems and the need for changes in light of
   i. Changing legislation, regulations, and best practices
   ii. Changing expectations and requirements of stakeholders
   iii. Changes in the organization's products, services, or activities
   iv. Changes in procedures, tools and equipment, maintenance, facilities, or management
   v. Changes to the organization structure
   vi. Advances in science and technology
   vii. Lessons learned from nonconformances, incidents, and accidents
   viii. Market preferences, changes, risks, and opportunities
   ix. Reporting and communications, including the effectiveness of procedures, training, and work center supervision
   x. Feedback (especially customer or employees)

PROGRAM AUDIT ATTRIBUTES

Periodic audits and inspections are necessary to evaluate program effectiveness and to reduce risking facility–managerial complacency. Roles and policies must be defined and evaluated as part of the audit process. Involvement of employees in the assessment process is critical to the success of the audit process. Managers and employees will need to be interviewed, and supporting documentation must be evaluated to determine the effectiveness of process or system implementation. Audits could include the following.

- Cross-facility audits (regulatory, program, or management system)

- Regulatory compliance audits
- Third-party and certifying agency audits
- Management system audits
- Financial or due-diligence audits

Senior management should review audit inspection and task analysis processes at least annually. The reviews should include the follow-up on proposed nonconformance corrective/preventive actions and should be done by using the basic criteria listed below.

- How well was the inspection or audit planned, and were checklists or written audit plans established?
- Determine if SOPs are being used and followed?
- Does the audit or inspection report address critical tasks or QEH&S concerns?
- Are the critical operations, tasks, and pieces of equipment identified and operating effectively and in conformance with customer requirements or design specifications?
- Do the reports identify corrective/preventive actions and assign implementation responsibilities, including the establishment of action completion target dates?
- Does the report assign and have documented follow-up procedures?
- Are risks effectively evaluated?
- Is training adequate?
- Are procedures periodically reviewed to ensure that they are still effective?
- What is the frequency of audits, job observations, and inspections? Is this frequency adequate given the level of risk associated with critical processes?

AUDITS, MEASUREMENTS, INSPECTIONS, OR TESTING ACTIVITIES

Independent management audits of the QEH&S program should be conducted at least annually. Objective evidence and documentation of program conformance is an essential part of the management system, and should cover the following.

- Determine whether QEH&S plans have been implemented and achieved
- Provide lessons learned from QEH&S management system failures, including accidents, near-misses, and rework
- Promote implementation of plans and risk controls by providing feedback to all stakeholders
- Provide information that can be used to review and, where necessary, improve management system performance or execution
- Track actual performance against goals and target dates

INTERNAL AUDITS OF QEH&S SYSTEMS

Scheduled audits should be performed to verify compliance with policy and to determine policy effectiveness. Competent persons, independent of the areas being audited, should be utilized to perform management system audits. Audit reports should be distributed to management and other stakeholders, which summarize nonconformance or process deficiencies and their prioritized corrective/preventive action plans.

Audits are a necessary tool to monitor key operational activities associated with the delivery of quality products and services and EH&S performance. Other types of management review can include periodic inspections or job observations and regulatory, technical process, or financial audits to determine conformance with business requirements.

Experienced, competent managers or professionals should perform audits by using defined and approved audit criteria. Mangers responsible for the nonconformance should be identified to get their concurrence on proposed schedules for recommended corrective/preventive actions and mutually agreed-to target dates for interim and long-term conformance. Auditors working with operations personnel should also attempt to determine the multiple root causes of nonconformances when designing or recommending corrective/preventive actions.

The frequency of audits and the audit criteria, including reporting and documentation procedures should be defined in the management system manual and other implementing documents. These procedures should also define the auditor's qualification, the process for selecting auditors, and the establishment of a process for implementing needed changes to policy, procedures, and equipment. The results of audits should be reported to senior management.

Auditors should use defined protocol to ensure audit process consistency and to perform audits objectively and impartially. Audit frequency should be determined based on the review of past audit results and the probability or severity of potential risks in the work areas, work practices, or activities. Audit plans must be tailored for each organization, and in some cases scored or weighted audits may be valuable in helping management prioritize actions.

Documented procedures should be established and maintained and should include the following details.

1. Allocation of resources for the audit process, including personnel and management time, materials, technical support, financial support, and training
2. Methodologies for conducting, documenting, and follow-up of audit results, including audit plans, checklists, questionnaires, interviews, measurements, and direct observations; the procedure should also define the process for communicating audit findings and recommendations to responsible managers and other stakeholders
3. Meaningful, open participation of all involved personnel should be encouraged; employees and supervisors should be asked to cooperate fully with the auditors and to respond to their questions honestly

4. Typical reports should include the following information:
   - Specific system or process deficiencies and the multiple root causes of those deficiencies
   - Recommended preventive and corrective/preventive actions or improvement opportunities

Although not required, periodic (as often as possible) job or task observations monitoring should be established to ensure that personnel are conforming to critical job requirements. This process must include appropriate feedback to stakeholders, so that desired behaviors and processes are reinforced and correction/prevention of nonconforming activities are initiated. If monitoring or inspections require collecting samples, then sample collection or analytical devices must be maintained and calibrated in accordance with manufacturer or regulatory requirements and/or internal policy.

Records of all calibration and maintenance procedures must be documented, maintained, and either referenced or included in reports by using the data established through the use of the calibrated device. QEH&S policies should also define the criteria for selecting measurement and test equipment. The criteria should define accuracy, precision, and environmental requirements necessary to collect representative samples.

Initial baseline or benchmark audits are important, as they will serve in the following ways.

- Help to define plans for the execution/implementation of management system policy and procedures (Gap analysis)
- Review compliance with legal requirements (legislation)
- Compare existing QEH&S with industry standards (best practices)
- Identify significant process risks
- Review past experience (statistics), previous audits, customer complaints or records of rework, product or service rejection, accident trends, environmental liabilities or releases, and permit compliance and also review current resource allocation for QEH&S (money, time, equipment, personnel, etc.).

Organizations should regularly review their QEH&S management system with the objective of continual process and performance improvement and compliance with relevant standards, guidelines, and regulations. The multiple root causes of nonconformance must be identified if process problems are to be eliminated, and if the lessons learned from these process evaluations are to be shared with others who are or could be affected by similar process or system weaknesses.

A defined process must be in place for the risk manager, quality manager, and EH&S manager to integrate reports combining the status of the continuous improvement process for their respective management areas. These reports could include the results of audits, job–task observations, and identified nonconformances and their multiple root causes. Corrective/preventive action plans should also be developed from the management review process and should be reflected in process or policy changes supporting the continuous improvement process. Process change, including changes to SOPs, tools, and equipment, must be reviewed and concurred to by the stakeholders, including operations, quality, and EH&S managers to ensure that conflicting requirements or controls are not established.

Organizations should maintain a statistical loss and accident–incident analysis procedure to measure performance. Data sources could include the following.

- Customer complaints
- Process nonconformance, rework, or process failure reports
- Environmental permit or release reports
- Injury–illness reports
- Insurance reports
- Incident reports
- Regulatory inspection reports or citations
- Industry reports
- Process behavioral-based job/task observations

Loss or rework data should be benchmarked internally and measurement should be developed by using frequency and severity rates losses per cost or as a ratio of cost per unit of production. In addition, analysis of accident–incident or nonconformance reports should be used to help establish program procedures and corrective action plans or goals.

There are many ways for management to periodically review the effectiveness of the QEH&S continuous improvement process. Every organization's approach will be a little different based on the complexity and risk associated with its operations. However, at minimum, audits will be required to effectively assess process execution. Some of the other measurement and documentation tools include using statistical process improvement tools for charting and tracking operational performance and can include some of the inspection and job–task analysis or observations procedures discussed in Chapter 7.

The goal of the management review process is to ensure consistent leadership and direction, including the provision of resources in the form of staff time, money, and tools to execute the integrated improvement process. Without senior management commitment and periodic review and direction the system will be doomed to fail!

THE ACTION PLAN

- Establish a documented system for QEH&S management system review
- Establish third-party and cross-facility or internal QEH&S audits
- Establish and follow-up on corrective/preventive action plan items

- Establish a system for integrating the following for management review:
  a. Status of previously agreed to actions, including those from risk assessments, audits, and nonconformance investigations
  b. Critical performance observation data
  c. Statistical process control, quality/rework data
  d. Systems design, readiness evaluations, and process improvement recommendations, including those from PETs

10

Conclusion

Try not to become a man of success but rather try to become a man of value.
—*Albert Einstein*

This book provides a desirable road map for integrating your QEH&S management systems. It also provides you with some risk assessment and monitoring tools for evaluating your existing system and its progress. Organizational management must understand that the process inputs or resources assigned to tasks are the keys to correcting/preventing substandard conditions. This system is important because, in most cases, substandard conditions, poorly designed or executed inputs are the precursors to substandard behavior or performance, including accidents.

This book has its foundation in my 25 years of environmental health and safety (EH&S) process management experience. I have observed that the most successful companies keep it simple. At various levels, they integrate quality and EH&S into a harmonized management system for process control.

It has been my observation that successful senior managers are leaders. They recognize that their success depends on taking care of both their internal and external customers. To accomplish this goal they are sensitive to the need for effective communications, and they recognize that as senior managers their internal customer is oftentimes their staff. I urge managers to ask, "What are we doing to ensure that our internal customers and staff have the needed resources to get the job done as designed, on time, within or below planned costs, and without injuring employees or contaminating the environment?"

Deming, Juran, and Crosby, ISO, and all the iterations of total quality management (TQM) and process management have their roots in some very common process management components. It's the goal of this book to help managers, supervisors, and some process technical experts better understand the value of a harmonized quality management system approach to total process management. A goal of this book is to help companies establish a process that considers and integrates process performance, quality, and EH&S.

We can significantly improve our quality and bottom-line performance by simplifying the process and looking at our management responsibilities differently. The views of process management, observation, follow-up, and reinforcement in this book are not necessarily unique except in the context of process and management integration of some process control tools. This approach to process management can significantly improve quality performance, including environmental health and safety.

I advocate a management commitment to harmonize the growing number of systems into a single management control process. Although some organizational guidelines, process flow charts, audits, and systems for accidents or nonconformance evaluations have been provided in this book as attachments, I emphasize that these guidelines are simply tools to be used by competent managers who understand this integrated approach to process management. The ultimate goal is to have a process enhancement team (PET) with heavy employee involvement in every work center and to have them share the best tools and resources that are available and have been developed for quality, environmental and safety management.

Benefits: This book should help operations managers, quality engineers, safety managers, industrial hygienists, and environmental engineers see the power of working together with the workforce to improve process performance at all levels. This goal is especially important for global operations that demand consistency and must recognize specific regulatory and cultural needs. Outside the United States process management is becoming very important for companies both in Europe and in countries with developing economies.

This observation is supported by the development of requirements such as the United Kingdom's BS 5750 (now ISO 9000), BS 8800 (OH&S Management System), and BS 7750 (now ISO 14000) standards on quality, safety, and environmental management. There are often other international standards that are consistent with ISO 9000 and 14000. Australia's safety map system is built on an ISO 9000 foundation, adding to the evidence for systems integration to support the continuing logarithmic growth of ISO certified management systems. There is also the draft OH&S standard 18001.

In the United States we are beginning to see increasing interest on the part of Environmental Protection Agency (EPA) and (OSHA) in management systems. Even the Chemical Manufacturers Association (CMA) has been evaluating the best way for its members to integrate its Responsible Care policy guidelines with these systems. All of these efforts reinforce the need for integrated or harmonized systems. Management, especially in this era of re-engineering, is always looking for ways to integrate functions and processes to improve operational and bottom-line performance.

By providing managers an introduction to total process management, with its roots in Deming, Juran, Crosby, and ISO management systems, I ask the reader to think about integrating quality and environmental health and safety programs in their facilities. I believe this approach is possible by using ISO 9000/14000 as potential models to build the integrated system or management process.

Chapter 2 reinforced the arguments for integrated management systems and focused on safety management, which until recently has not received the attention it deserves in the ISO management standard-setting process. Introduction to measurement and behavioral observations for quality and safety were also introduced in several places in the book and discussed in some detail in Chapter 7. The reference section of this book identifies additional resources if you are interested in a more in-depth discussion of performance management and behavioral safety. The discussion in this book is simply to show how these approaches can be extremely powerful measurement and observation tools for an integrated process management system.

The book repeatedly emphasizes the roles of senior managers and supervisors. It also discuss the policies, goals, and objectives that must be established and followed-up continuously as part of the total process management process. Planning has been discussed as the foundation for designing and successfully implementing a total process management system. Clear roles and responsibilities must be established, as well as a process for systematically reviewing process/resource inputs such as the procurement process, training, maintenance, and procedural controls.

The concepts of multiple root cause analysis, risk management, risk mapping, process nonconformance, cost benefit analysis, and systems safety have been introduced and discussed in several sections of this book. These steps are the first in effectively evaluating the level of corporate and process risk and determining the necessary and most cost-effective controls to eliminate or minimize the risk.

Measurement and evaluation are discussed as critical elements of the total process management process. Audits, including initial Gap analysis assessments and managerial, technical, or process, regulatory, cross-facility, third-party, and follow-up audits are discussed. On-going systems evaluation and performance improvement-reinforcement is encouraged by using tools such as performance management or behavioral safety tools or other critical task-process observations procedures. Tools such as job–task analysis and follow-up on corrective/preventive action or process implementation were discussed as part of the management review and continuous improvement process. The necessary elements of and the implementation of an integrated total process management system include the following.

- Employee involvement
- Training
- Communications
- System documentation
- Records

- Document control
- Design control
- Purchasing
- Hazardous materials/waste management
- Management of contractors and vendors
- Emergency preparedness and contingency planning processes
- Use and maintenance of equipment, tools, and the facility

Operational control as discussed in this book has placed a lot of emphasis on the specific use of procedures and on controlling the work environment at the process resource input and design stages. Part of the overall management control process is the understanding of the multiple root causes of process nonconformance. This knowledge is critical for understanding how to improve the process at the design or resource input stage and to ensure that substandard conditions are not designed or introduced into the process and becoming the precursors for substandard performance and process failure.

Critical to mastering process improvement and accident prevention is understanding the management system and the elements of the system that cause process nonconformance and incidents. The concept of multiple root cause analysis and the tools for process nonconformance or accident investigations was discussed in detail in Chapter 8, including the need to document the total costs of nonconformance (direct and indirect costs). Until management understands the financial impact of substandard process design/execution and the benefits of process change communicated in financial terms such as discounted cash flow, return on investment, and payback period, only stop-gap measures will be applied to the process enhancement or corrective/action effort.

Identifying the causes of process nonconformance and sharing lessons learned to correct/prevent a reoccurrence are some the most important tools supervisors and managers will have on the path to continual improvement. If you can recognize what went wrong and why without blaming employees, you can improve the process and prevent a reoccurrence. Blaming fails to identify the multiple root causes of the problem that exist in the system and is the lazy, uniformed manager's way of dealing with process problems.

Management reviews and defined processes for measuring performance have been discussed briefly as essential elements of a successful total process management–improvement system. Chapter 9 discusses some of the tools managers can use and, like Admiral Rickover, I believe that audits and the follow-up of corrective actions are critical to successfully implementing any management system. The appendices of this book contain various management and technical audits, which can be edited or merged as appropriate to develop your organization's audit tool. These tools are presented only as examples. I would encourage you to edit or merge them to more accurately reflect the needs of your existing or proposed management system. The enclosed audits are provided only to help you get started. To make these audits more useful, you will need to carefully edit and merge check sheets, filling in critical gaps so that the audit makes sense for your organization and considers regulatory requirements in your country or state.

To be successful, companies must accomplish three tasks.

- Maximize revenue generation and market penetration
- Control expenses, especially overhead
- Prevent all forms of loss as evident in injuries, quality, rework issues or environmental liability (Losses directly reduce the company's bottom line!)

I do not see a clear distinction between overhead and variable cost models for consulting verses a Macy's shirt shop or a manufacturing operation. They are just different. Profitability requires that managers focus on revenue generation and controlling their costs, especially losses. However, in some cases managers do not really know what the costs are for providing various services and products, or they don't have a complete understanding of the losses resulting from environmental health and safety or rework. These costs are therefore not fully recognized or charged back to the appropriate work center.

The discussion of life cycle risks and documenting direct and indirect costs was provided so that managers could begin the process of more accurately recognizing and controlling *all* costs associated with products and services. I have seen consulting and production organizations continue to keep people busy but not with a fee for service, work, or efficient production that allowed them to pay their costs and make a profit. In fact, in some cases they begin the process of buying work or customers and the slow spiral into negative profitability.

This lack of understanding can be especially true at the supervisor level or with technical consultants who do not have any financial or management training. In consulting we oftentimes use salary multipliers and utilization goals as models to help staff and supervisors understand how to build profitable projects and avoid buying work. In production or other types of service operations you will need to use similar easy-to-understand ratios to ensure that supervisors have a prospective forward-looking tool to monitor their progress between financial statements.

The validity of an average multiplier for an office is good only if the office meets its sales/revenue goals, which is again like Macy's selling shirts. In the four states that I had offices in and with our overhead structure, I knew I could make our profit margins with an average 3.0 multiplier, as long as each office averaged 80% utilization of available hours or 68% of total hours. To ensure that we maximized our utilization in each of our offices, managers met with the project managers and staff weekly. The purpose of the meeting was to look at our projects, push the schedules as much as we could and share the work with staff who were underutilized (whenever possible).

We have some control over fixed costs like salary and rent, but, for the most part, once set we need to live with them unless we plan to fire staff. Rent can be controlled to some degree, and most variable costs can or should be project or business-development related.

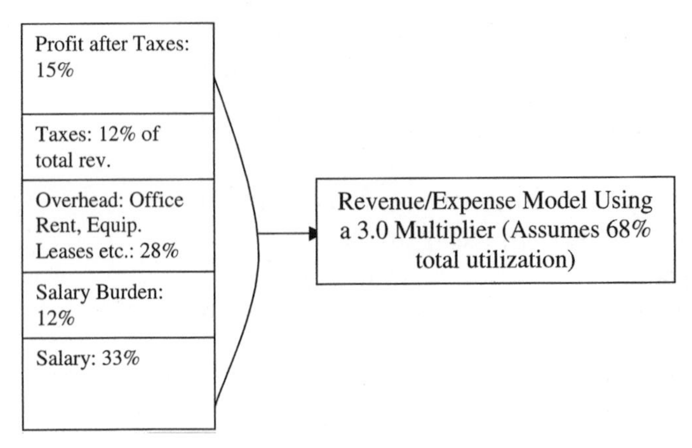

Figure 38 *Consulting Multiplier/Utilization Cost/Profitability Model.*

The biggest opportunity to control costs once we have minimized overhead is through project or production management. In many instances, employees and supervisors do not do a good job documenting time and expenses on a project; this instance is especially true in many consulting operations. The opportunity to maximize revenues and profitability takes place when the product or project is planned or designed. If costs, product design, and the production process are not fully recognized, unnecessary costs, process risks, and inefficiency will be built into the process with all of its associated liabilities and shrinkage of the bottom line.

The production or project plan gives the manager the opportunity to recognize and capture revenues for changes in the scope of work, product, or service design. The plan is also intended to ensure that all planned time and expenses for the job are recognized and included in product or service pricing. The product or project plan also creates a smoother more efficient project or production process that increases the potential for profitable service delivery. Good project and product design for QEH&S may increase the product or service cost 0–5% more, versus an increase in costs of 10–20% if the product or design must be retrofitted with controls.

The use of good systems to fully recognize the costs of service delivery or production and life cycle costs are critical. In some instances, we may decide that we need to increase prices or no longer produce a product or provide services because it can't be done profitably at current prices.

Before going down the path of implementing a management system, I also recommend that an organization take an honest assessment of itself to better understand its existing culture and willingness to make some of the management changes discussed in this book. Keep in mind that many substandard behaviors are the result of the environment and the substandard conditions created by the existing management system.

Culture: Discussed below are some of my views about management leadership and the philosophy of managers in creating the work environment that will either successfully support process change or possibly make things worse. Making things worse is a risk of an apparent disconnection between the head of the dragon and the rest of the organizational body, which is especially apparent in negative management environment or skeptical culture that is not ready for it.

Described below is the type of culture I want to work in and that I believe most people want to work in, and, more importantly, contribute to. Before you can initiate systems changes or implement behavioral based systems, you must first examine the culture of your company at all management levels and in all departments. If you don't do this you run the risk of making things worse by not truly understanding the beliefs and myths that create the daily reality for the people who will be affected by the proposed changes. Also keep in mind that change management is critical; even positive change will be disruptive at some level and should therefore be anticipated.

Implementation of a management system without a cultural review and audit is kind of like a physician making a diagnosis over the phone and thus prescribing the wrong medication or a medication that reacts with other medications and makes you more sick, or worse, kills you. The head of the dragon must determine if the organizational beliefs are consistent with the continual improvement process and the change in direction it wants to take. If there is a serious disconnection, management must first look at itself, its integrity, and the consistency of its communications and actions.

If the organization is not ready, you will need to evaluate how you will make the change in culture over time or modify the system implementation. How can the changes be made with permanence and integrity if it is not real for the leader or the supervisor implementing it? The commitment and involvement with these beliefs will be superficial at best.

Behavioral-Based Programs should not be implemented unless the organization can affirmatively respond to the points listed below.

- Will QEH&S or systems safety or process improvement be supported and effectively implemented?
- Do people understand the process and know what to do in the event of a nonconformance?
- How does management review, evaluate, and implement continuous improvement and how involved is management in creating the necessary critical focus?
- How is decision making shared?

The bottom lines is that everyone must be working towards meeting the customers needs and sharing the responsibility for customer relations, marketing, and business development. Employees need to understand how they contribute to the overall goals and objectives of the organization and to share in the rewards they have helped to generate.

I discussed the Pygmalion effect in Chapter 1. Is your organization a positive or a negative Pygmalion? What are your beliefs and expectations about people, especially those beliefs that center on age, race, ethnic, or regional orientation? Do you believe people are lazy or hard-working, honest or liars and cheats? Oftentimes the universe will align itself to support your beliefs; so who is at fault in a culture or management system that expects the worse? Good attracts good, and positive attracts positive.

- **Integrity and Ensuring a Fearless Environment with Trust/Mutual Respect:** Integrity or simply "being your word", living a life that makes your co-workers, other managers, staff, and customers know they can count on you. If you commit to getting something done they know it will be done. No question about it! It also means that as a leader you will create an environment that stamps out fear and allows open, honest opinions and other communications to be shared. Without this kind of culture, corporate or process enhancement will not be possible. Your most valuable resources, especially the first-line supervisors and managers who are the heart of the dragon, must be supported. They are the ones who give the organization life and who make it happen day in and day out.
- **Ethics:** Honesty and integrity are the backbone of any successful relationship. Dishonest or unethical behavior within an organization or with clients must be dealt with swiftly. The fastest way for an organization to fail is to loose its credibility and/or to be perceived as unethical or dishonest. This standard must be absolute for everyone in the organization.
- **Leadership:** Organizations increasingly need hands-on management. To build strong teams with higher quality and profits, managers must be willing to roll up their sleeves, team with other managers, and work alongside the staff to motivate them to excel as partners in the business. Leadership is needed to define the process, make the system work effectively, and ensure that resources are allocated to the process. It is here that a distinction should be made, and that is the difference between a manager and a leader. Webster's dictionary defines a *manager* as "one who conducts business or household affairs", and a *leader* as "a person who has commanding authority or influence." Which are you?

The truly great leaders in history had strong charismatic personalities and visions that they truly believed in. Leaders make you want to be part of the vision. There is an excitement and sense of accomplishment that comes along with partnering or working for a leader, such as the level of excitement and challenge that John F. Kennedy created when he challenged Americans to send a man to the moon or to sacrifice and offer their skills to a new organization called the Peace Corp. To accomplish the difficult tasks before them, leaders must be the respected and experienced people who know their business and their customers. They are innovative, creative, goal-oriented people with very good communication skills. They can define and share their vision, inspiring others to "buy in".

Because they understand that they have both internal and external customers they can work astutely across organizational lines to accomplish or create shared goals. Leaders can get others to accept new ideas and obtain organizational support for them. Leaders respect themselves and those around them. They are not weighted down with false pride and, most important, they have integrity. A leader is someone you can trust, and who is loyal not only to the shareholders and other senior managers but to the people producing the goods and services for the customer. Their word is their bond. Their word means everything. You can take their promises or commitments to the bank! They are energized and provide feedback and direction to ensure that the process works, including its continual improvement.

Leaders also have the courage to make difficult and sometimes unpopular decisions. They are committed to their integrity and doing what is right, while knowing that in many cases they will not have 100% consensus and at times may even be in the minority with their beliefs. Leaders have the courage to do what is best for their companies, their customers, and their employees and not to do simply what will give them the greatest short-term gain.

Leaders oftentimes see those people who work for them as their internal customers. As such, they are constantly looking for ways to provide them with the tools and resources needed so that they as a team can provide quality products and services to their external customers on time, within budget, and with a reasonable profit margin. Leaders also have the courage to develop their subordinates, or to even plan potential succession, with people fully trained and ready for promotion in the event of a contingency or change in business demands or organizational structure.

When I chaired the management enhancement committee at Mare Island, I visited Johnson & Johnson to review its quality and management development systems. It was interesting to see a company that was so well-organized. One of the more interesting aspects of their system was their commitment to succession planning.

Like the presidency of the United States, there must be a clearly defined path of authority for leaders to step up to in the event of an emergency. I find it interesting that the framers of the Constitution saw this system as important and that many so-called modern businesses either ignore this need or keep *managers* who are afraid to be leaders (with a ready successor working closely with them!). Leaders should be mentors who value the strength of a planned succession. They know that it strengthens their position and the organization. It is this self-confidence that truly distinguishes the difference between leaders and managers.

Six Sigma, which was briefly discussed in Chapter 7, is different in that the companies that follow this practice recognize that their manager/leaders are critical to the success of the process. Management becomes actively involved committing 15–20% of their time to the process and in developing Six Sigma Masters, or Black Belts to lead the process. Rather than delegating everything down the line and forgetting about it between reports, companies implementing this system are totally committed to the process. About 10–15%

senior management time focuses on the quality management system and the Six Sigma leadership processes. Without it, Six Sigma or any other management system will fail. Joe Welch, CEO of General Electric, requires this level of commitment to process mastery, which is a prerequisite to becoming an executive in his company.

Management Control: Managers are resources for their staff and clients. They must meet and work continuously with them and their clients to hear and understand what is wanted or needed. It is management's job to remove obstacles from the process path. They serve their internal clients (staff) and managers to ensure that the job is done on time and in accordance with specified requirements and budgets. To accomplish this goal, management must clearly understand the work process, including administrative and financial needs. To facilitate this process understanding, tasks must be broken into steps and evaluated for recognizing potential problems and correcting existing ones. It is management's job to generate continuous improvements and reduce rework and accidents.

Management control means that project and quality management contributes directly to the bottom line. Profit and loss management is not simply a financial management issue. Cost avoidance, revenue and contract maximization, eliminating rework, and accidents are critical in today's competitive markets. Managers must be able to develop key personnel and be able to delegate responsibility. Every manager should be developing potential replacements to avoid business interruptions when they are gone and to provide for a smooth

Figure 39 *Controlling the Dragon.*

transfer of authority if they should be promoted or transferred to another job.

Training/Staff Development: Business and technology change very quickly today. Public schools are not adequately supported by the business community or government resources. The end result is that management will need to be more aggressive in evaluating staff development and training needs or face obsolescence and loss of productivity and market share. Training is a vital element in the continuing quality improvement process.

New processes, products or services will require staff development and training to be effectively designed in the first place. Once a company has a new product, service, or process for executing a particular task, staff training will once again become important at some level depending on the complexity of the tasks or associated requirements.

In addition, as Deming pointed out, staff training and development are investments in your most vital resource. Analogous to fertilizing the soil to create the environment for growth (new idea, new products or services, or better more cost-effective ways to deliver existing products). As companies focus on increasing their pre-tax operating income, training and development too often becomes the first expense to be cut. If a business is struggling to survive, some cutback may be necessary. However, if the cuts in training budgets are too deep or prolonged, you are mortgaging the future of the business or temporarily postponing the "last rites" for a business that is essentially dead.

Compensation Practices: Compensation packages and bonuses must reflect both the needs of the employee and the contribution/value of the employee to the company (company profitability). Compensation that is competitive with equivalent positions or professional classifications within a defined region is important, and the employee's salary and bonuses should be fairly distributed. Reference 5 discusses the increasing turnover rates with many companies as employees sense less loyalty on the part of the company and perceive an opportunity to significantly increase their salaries by joining another company, often times with sizeable signing bonuses.

Shareholders need to realize a significant return on their investments, but as companies down-size, merge and consolidate operations, the gap between the executive's pay and the people generating the revenues has increased significantly. The average employee's salary has increased 27% over the past decade, whereas executive pay has increased 163%. During the same period, the S&P increased over 218%. Reference 5 illustrates that the increasing salary gaps and lack of loyalty have created opportunities for significant pay increases just by quitting a job and taking advantage of our current full-employment economy. Benefits realized by companies will be wiped out by these turnovers as the loss of personnel, talent, and replacement costs, especially at higher salaries, wipe out current short-term gains realized by the executives who have personally benefited by the disparity.

Employee Evaluations and Bonuses: Employee evaluations and the criteria used to measure and evaluate performance must be clearly and easily understood.

It also must be fair and as objective as possible (few systems are totally objective). Quality, productivity, dependability, and cooperation are critical job performance elements. Bonuses should be based on individual performance and the performance of the organization as a whole. It is better to have committed partners in your business, all watching overhead and rework, than a business filled with staff putting in a moderate effort from paycheck to paycheck. As mentioned earlier, I encourage you to become familiar with Lincoln Electric's bonus system — which is discussed in several places in this book, including Appendix III — or SAIC, the largest employee-owned company in the world (see Appendix V). SAIC is especially significant when you consider that they are a 30-year-old science and technology company.

EEO: Everyone deserves equal and full participation in what society has to offer. Preferential treatment or biased and unfair treatment because of gender, age, religion, race, or sexual orientation cannot be tolerated. People should have equal opportunities, but after the door is open advancement and additional opportunities should be based on performance and ability to contribute. Harassment of any kind, especially intimidation or sexual advances, cannot be tolerated. The workplace should be a safe environment that people enjoy coming to and being part of. EEO is a management philosophy that supports that goal.

Labor–Management Relations: The ability to cooperate, to communicate effectively and actively, and, most important, to listen is critical to anyone's success. Most labor management problems can be resolved if mediation or negotiation is supported with effective communications and a sense of ethics and fairness.

Business Development, Marketing and Lobbying: Managers must have experience in business development and be able to see the value of interacting with and supporting their communities and clients. Management must understand changes in the market and the needs, wants, and desires of their clients or prospective clients. Competition is not bad in that it can hone the skills or the quality of the services or products being offered. Sometimes, cooperation, collaboration, and joint venturing can reap much greater rewards with much less risk, effort and greater profits than killing each other off in the marketplace.

It is usually more powerful to look at how best to serve the needs of the customer, which may then dictate cooperation or even teaming among competitors. Lobbying and working with community groups or leaders is essential to facilitate some changes and to be able to better care for the needs of the customer. Creativity and the ability to generate new possibilities, exercising technical or hands-on managerial experience are essential for organizational success.

Community Support: No business is an island! Community support in today's environment is essential, especially with more small businesses. Many schools are producing functionally illiterate high school graduates, while violence and drug abuse are becoming a reality in every community. Businesses interested in long-term success must provide leadership and resources to support the community at large, especially something as critical to the dragon's long-term health as is education of our kids as future leaders and employees. Leadership requires a commitment to understand and manage the needs of the organization and to balance those needs and resources with those of the community supporting the business. Education, health care services, and safety are essential tools to end violence and drug abuse. Business leaders must make this a common theme in their businesses and the community they reside in. Mutual support and collaboration are needed.

CHANGING YOUR BUSINESS

Chapter 1 has a process flow diagram model for implementing a management system, including periodic management review and continual improvement based on Deming's "Plan-Do-Correct" model for continuous improvement. This process flow diagram may seem overwhelming at first glance, but in reality it simply clarifies Deming's model. As mentioned earlier, the process cannot be executed overnight, and the critical steps in the process must be implemented over time based on your prioritized risks and corrective/preventive action plans. The model is far from being chiseled in granite. It simply identifies issues and tasks that must be considered when designing or implementing your management system.

To be successful, however, management must have continuous monitoring and feedback as required by ISO and most effective management systems. Chapters 7 and 9 should go a long ways towards establishing the management review and continuous improvement processes. Most important, I recommend that managers follow the "KISS" principle and the prioritized process change initiated with your risk assessments. Monitoring for management feedback can include audits, job–task observations and or performance management or behavioral safety programs.

Companies must involve everyone in looking at ways to cut costs and enhance performance. In addition, targeted processes should have process enhancement teams assigned on a prioritized or risk-ranked basis. The purpose of these teams is to define and understand the critical process and to eliminate or control the risk and improve efficiency. Communication in the organization must be open to including the sharing of financial information and lessons learned from process nonconformance. Individual development, including cross training and succession training, is critical for the long-term health of an organization.

Understanding and managing risk is one of the most important aspects of the management system's continuous improvement process. However, companies cannot be successful if they are so risk-adverse they fail to take advantage of business or market opportunities. I must now create a distinction for you between operational risk management as discussed so far in this book and business or opportunity risk. Although related, they differ. You can even use some of the risk assessment tools for operational risk to evaluate market or business opportunity risk; however, to have a company that will grow you must have an appetite for business risk. Without it, growth is impossible or at best severely frustrated!

Business leaders must also be the ultimate decision-makers for business or opportunity risk decisions. They are the ones who will lead the organization not the CFO or legal council, who have a tendency to be risk-adverse and to perceive risk in everything a business does. You need these advisors, but you must be able to put their council in perspective and not regard it as the ultimate decision-maker in market or business opportunities. Too often managers fail in their leadership capacity. They look to others to make these critical decisions. Others retreat into various forms of analysis paralysis that will result in a negative decision with regards to the business opportunity, while the entrepreneurs and business leaders who are quicker on their feet take advantage of the emerging opportunities. This form of leadership failure will leave your organization with lost market share and costly attempts to regain a leadership position in the marketplace. Keep in mind that it costs 5–10 times as much to gain customers as to retain them, and of those customers who leave, only 10% will ever return.

Business is not without risk, and the rewards would be very small if it were risk-free. In this book, I advocate distinguishing between operational and business opportunity risk. I believe that operational risk must be very clearly understood, prioritized, and controlled, as every cent that is saved by avoiding process nonconformance will go directly to the bottom line, increasing pre-tax operating income.

However, a willingness to take calculated business opportunity risk increases the revenue line and establishes a foundation for growth and a platform that will allow companies to be truly responsive to the needs of their customers. Therefore, operational and business opportunity risk are not mutually exclusive. They must both be considered with appropriate tools applied to understanding and controlling them. It is the role of the leader to manage this process and thus ensure the future of the organization!

I hope that this book will be useful to all managers and organizational leaders who are trying to understand and control risk in their business. I hope that in time I will be able to meet with the readers of this book to discuss your experiences with integrating management systems and to help you with your process and some of the tools discussed in this text.

Appendix I

Audit Attributes

POLICY AND MANAGEMENT COMMITMENT

References: ISO 9001: 4.1; ISO 14001: 4.1, 4.2.4

4M1 Top management has formulated a quality and environmental health and safety (QEH&S) management system policy consistent with the facilities, potential quality, operational, EH&S risks, and environmental impacts of its activities, location, products, and services. Objectives and targets are established and followed-up to ensure system execution continuous process improvement and risk reduction.

- Does the management system policy address accountability in terms of allocating resources, assigning and executing commitments, and setting objectives and performance targets to maximize performance and eliminate environmental health and safety liability while enhancing quality and process performance?
- **Note:** Interview employees at various levels to determine if they can describe their individual role in the QEH&S process. If the interviewed employee is part of the management, can the system effectively describe lines of responsibility and authority and who he/she interfaces with if issues potentially impact across more than one department?
- Is the policy integral to the organization's overall mission statement, vision, core values, management activities, products, and services?

- Is there a copy of the written and dated policy document that clearly states quality and EH&S objectives and commitment to process improvement/risk reduction?
- Are QEH&S policy statements that are issued from an operations management level high enough to ensure effective implementation?
- Where required, have specific policies/procedures been developed for quality and EH&S process risks?

4M2 Are the costs for accidents/incidents, injuries, or illness (including fires, personnel injury, environmental or process nonconformance, HM/HW [hazardous materials/hazardous waste] spills, compensation, penalties, etc.) charged back to the work or profit center?

- Are there incentive or recognition programs for quality enhancement, injury/illness, disability, or sick leave reductions? If so, please describe and provide documentation.
- Are there medical case management systems or return-to-work systems in place for injured or disabled workers? If so, please briefly describe and provide documentation for all levels of the organization, including contractors and vendors.

4M3 The quality and EH&S plans allocate resources and establish prioritized, measurable organizational objectives for process improvement. The plans must include prioritized target dates and responsibility for all committed actions and establish measurable organization objectives for process improvement for all levels of the organization, including contractors and vendors.

4M4 Include in the management policy a commitment to continuous improvement, more profitable and efficient processes, prevention of releases of HM to the environment, and workplace, control and monitoring of all potential exposures to HM/HW, with a goal of eliminating potential exposures. Workplace occupational health and safety risks are identified, evaluated, and controlled.

4M5 Policy includes a commitment to regulatory and process conformance.

4M6 Document the quality and EH&S policy and ensure that it is communicated and understood by all employees and that it is available to regulators and the public (stakeholders).

- The policy has been developed by management in consultation with employee representatives and other stakeholders
- quality and EH&S performance is included in the organization's annual report or equivalent

Documentation/Records

EMS/QMS Policy

Organization Charts

Management Review Records

Policy

References: ISO 9001: 4.2; ISO 14001: 4.2

RESPONSIBILITY, ACCOUNTABILITY, AND AUTHORITY

References: ISO 9001: 4.1, 4.2; ISO 14000: 4.0, 4.2.1, 4.2.3, 4.2.4, 4.3.1, 4.5

4M7 Responsibilities for quality and EH&S are defined, documented, and clearly communicated, including assignment of authority, resources, and budgetary support necessary to implement and maintain the system. (Resources include financial, technical, tools, equipment, and human.)

- Has a senior executive of the organization appointed an individual by title, as the QEH&S management representative? Has this individual's responsibility and authority been conveyed throughout the organizational?
- Responsibility and authority for establishing, implementing, and maintaining the requirements of the QEH&S management system are defined, including legislated or regulatory requirements.
- Management personnel are held accountable for quality and EH&S performance within their individual work areas.

4M8 A manager(s) is assigned responsibility for the integrated management system conformance. Clear responsibility is established for risk management, EH&S management, operational, and quality management. These managers also report to senior executives. The organization obtains advice from qualified quality, process improvement, and EH&S professionals (in-house or external).

- Has a senior executive of the organization appointed an individual by title as the quality and EH&S management representatives?
- Are competent individuals or consultants retained and assigned specific technical and managerial responsibility qualified in the anticipation, recognition, evaluation, and control of quality or EH&S risks?

Note: Competency may be demonstrated through experience, education, training, or possession of a nationally or internationally recognized certification (e.g., certified industrial hygienist, certified quality managers, or auditors, associate risk managers, etc.)

- Is the relationship of this individual's responsibility and authority clearly defined with respect to all departmental, inter-departmental, or matrix responsibility and authority? Is performance periodically evaluated?
- **Note:** Interview these individuals to assess their understanding of their responsibilities. Do their responses mirror that of organization's documents?

Auditors may elect to interview other individuals within the organization to determine if they understand these individuals' responsibilities and if activities are consistent with defined roles.

- Do these technical management representatives participate in the development and implementation of policies and programs?

Documentation/Records

EMS/QMS policy statement and manual
Organization charts

MANAGEMENT AND STAFF JOB DESCRIPTIONS

References: ISO 14001: 4.1; ISO 9001: 4.

4M9 Document the organizational structure of the quality and EH&S management in a manual. The system should include process work instructions and other standard operating procedures (SOPs) to ensure conformance with QMS /EMS requirements, or regulations.

- Are these guidelines readily accessible to all employees?
- Do these procedures ensure identification and conformance with new policies, SOPs, legislative changes, laws, standards, and recent modifications to the process or service delivery?

Documentation/Records:

QMS/EMS manual
Operations procedures, SOPs
EMS/QMS plans, objectives, and target dates, including follow-up documentation

QUALITY SYSTEM

References: ISO 9001: 4.2; ISO 14001: 4.0, 4.3.4, 4.3.6

4M9 A clearly defined plan for executing quality and EH&S objectives and milestones has been established:

- Describe the process used to stay current on newly enacted federal, state, and local laws and regulations

IMPLEMENTATION AND OPERATION

References: ISO 9001: 4.2; ISO 14001: 4.2

Organizations should develop the capabilities and process improvement support mechanisms necessary to achieve their QEH&S policy, objectives, and targets. Policy should be clearly communicated to all managers, supervisors, and employees.

LEGAL AND OTHER REQUIREMENTS

References: ISO 9001: 4.2, 4.4; ISO 14001: 4.2.2

5M1 The facility has the resources and procedures in place to identify and access applicable legislative or regulatory requirements, including periodic reviews to ensure that changes are identified, understood, and considered in the planning process as appropriate.

OBJECTIVES AND TARGETS

References: ISO 9001: 4.1; ISO 14001: 4.2.3

5M2 Ensure that the management system policy establishes a framework for setting objectives and milestones.

- Is there objective evidence that reviews and revisions are conducted periodically, for example, committee minutes, data change drafts, or ERRATA sheets?
- Does the management system provide appropriate opportunities for stakeholder input in the development of goals and objectives (this includes employee involvement, e.g., quality or safety committee)?
- Are QEH&S goals and objectives communicated and understood throughout the organization by its stakeholders? Prevention of work-related injuries and illness, quality, and environmental liability must be a priority.

Note: Auditors can interview various employees within the organization to evaluate if goals and objectives are understood.

- Does the organization use QEH&S performance benchmarks to gauge its performance and continuous improvement efforts?

- Where required by legislation, designated work groups are identified and EH&S representatives are elected and trained.
- Competent personnel have identified and assessed process risks and controls.

5M3 Ensure that objectives are consistent with the policy statements generated by the facility's management system.

- The structure of the QEH&S work and risk assessment groups are established, employees are informed about the structure of the work groups and about how representatives are nominated and/or how they can participate in this process.

COMMUNICATIONS

References: ISO 9000; ISO 14001: 4.3.3

5M4 There is an established internal communications system for process enhancement that specifically addresses QEH&S issues.

- Are there minutes that indicate the QEH&S committee meets at least every three months to discuss issues and that the minutes are distributed throughout the workplace?

Note: Acceptable communications can be accomplished in many forms (bulletin boards, memos, formal meetings, etc.). In some cases, the form, content, and distribution of the communications may be dictated by law, in which case compliance should be checked at least on a selective basis. Review selected nonconformance reports and corrective/preventive action plans to determine investigation process effectiveness.

- Where designated work groups are established, employees are informed about the structure of the work groups and how representatives are nominated/selected and how they can participate via this process.
- Is there a list of employee and employer representatives on display or are there other methods of communicating this information to employees?
- Are there quality circles or similar grass root or labor management committees evaluating critical or high-risk jobs or processes? Committee(s) focus on risk assessments and the process and procedures to control risks.
- How are recommendations executed and followed-up?
- Does the organization encourage employee reporting of hazardous conditions and process risks without fear of reprisal?

- Are there written procedures for reporting risks and communications regarding corrective/preventive action, including follow-up and response to the employee?

Note: Review representative exposure communications to determine if the communication is provided promptly and is clearly written and understandable. Verify the method used by the organization that ensures that employees understand the nature and implication of exposure monitoring results or job-task observation. Interview selected employees to determine if organizational efforts to solicit and act upon employee comments are in place and genuinely encourage employee input. Ask to review cases where employee input has lead to corrective measures or QEH&S improvement.

5M5 Is there an established system for receipt of, response to, and consideration of external communications (e.g., customers, industry, community stakeholders, regulatory, etc.)?

- Procedures exist to enable stakeholder consultation regarding process quality or EH&S changes.
- How does the organization communicate appropriate biological, medical, or environmental exposure information to employees, and does the process comply with relevant laws, regulations, policies, and standards?

RISK ASSESSMENT PROCESS

References: ISO 9001: 4.14; ISO 14001: 4.3.6, 4.4.2

6M1 The organization should develop a documented process for both qualitative and quantitative operational risk assessment. This process must include input from affected stakeholders and should be used to identify and prioritize potential operational process risk, the multiple root causes of process deviation or nonconformance, and changes to the process to mitigate risk, including risk transfer.

- The quality and EH&S committee(s) focus on risk assessments and the processes or procedures to control risks.
- Competent persons have identified potential hazards and assessed the risks in the work process.
- Are hazard identification and risk assessment procedures documented (e.g., hazard inventory, consultants' reports, manual handling assessments, noise assessments, etc.)?
- Is there a document that requires the use of a hierarchy of QEH&S controls (engineering, protection equipment, administration) in the determination of control measures?

6M2 The facility has identified and evaluated the operational risks associated with its activities, products, and services, including risk to employees, the community, contractors, and the surrounding environment.

- An organization-wide strategic plan has been developed and implemented to control identified process risks, including the following.
- Specific plans associated with products, processes, projects, or sites have been developed
- Updated QEH&S quality control, inspections, and testing techniques
- Plans to measure performance through audits and system reviews
- Plans for execution of process corrective/preventive actions and measurements, resulting from incident/accident investigations or non-compliance reports to control operational risks or workplace hazards

6M3 Risks are prioritized by using a disciplined and documented process.

6M4 Significant risks are included in the facility's objective-setting process.

6M5 Information related to the reduction of identified risks, their ranking, interim controls, and objectives are established to reduce the risk potential to as low a level as reasonably achievable.

MEASUREMENT AND EVALUATION

References: ISO 9001: 4.10, 4.11, 4.12, 4.17, 4.20; ISO 14001: 4.41, 4.44

7M1 Are procedures in place to routinely conduct independent internal, cross-facility audits, and inspections of processes, equipment, or facilities? These procedures can (recommended) include critical job observations such as those performed with behavioral safety job observations.

- Does the organization have documented inspection procedures to evaluate the quality and EH&S, including follow-up status of inherently hazardous processes or activities?
- Are all known physical and work process hazards included in the inspections? Has there ever been a regulatory inspection?
- Has the insurance carrier or QEH&S or selected consultant inspected the site? Describe their findings or provide a copy of the report.

7M2 Management system policy specifies the frequency of monitoring or measurement activities and establishes a process for identifying significant processes or systems to evaluate, including the development and use of internal audit protocol.

7M3 All audits or inspections must be conducted by competent personnel, with their competency documented through training, certification, or experience. Do qualified auditors perform evaluations?

7M4 Procedures are established to ensure that audit and inspection criteria are current with regards to regulations, internal policy or industry standard changes. Do documented inspection and evaluation activities verify and conform to specified organizational requirements or regulations?

7M5 Inspect products or a service delivery process at appropriate stages of production or development to ensure that they meet design specifications and to control process QEH&S requirements. If a product does not conform to requirements, hold components or the service process until it meets specifications. Are there procedures to ensure that product servicing or maintenance is provided in accordance with specifications and to recall products if urgently released materials have a defect?

7M6 Incoming product is not used until it has been inspected or verified that it conforms to requirements in accordance with the quality plan and customer requirements. Level of controlled inspection depends on the level of control desired with the vendor or subcontractor. Incoming product can be released for urgent production needs prior to verification if it is positively identified for possible recall. Then, the finished products and service deliverables are inspected to ensure conformance to quality plan and customer requirements before being released.

7M7 Has the facility identified the need to apply statistical techniques to process control and product verification, including critical job or task observations to conform to quality or EH&S performance standards?

AUDITS, MEASUREMENTS, INSPECTIONS, OR TESTING ACTIVITIES

References: ISO 9001: 4.11, 4.12, 4.13, 4.17; ISO 14001: 4.4

7M8 Does the quality/EH&S management representative conduct routine evaluations, inspections, and audits, including documentation, reporting, and follow-up? The term routine indicates evaluations are conducted daily, weekly, or possibly less frequently if static working conditions are encountered but at a planned frequency.

- How are program or inspection deficiencies documented and followed-up to ensure corrections are in place?
- Are there documented procedures for inspections and audits, including schedules and checklists?
- Are all known physical and work hazards or process risks, included in the inspections?

- Are inspections documented?
- Does the organization document procedures to evaluate the QEH&S inspection and audits?
- Is there a schedule for inspection and maintenance of the facility plant and equipment, and does it verify QEH&S implementation of requirements legislation, standards, and codes of practice?
- Verify that the organization inspects and evaluates newly purchased or installed personal protective equipment, engineering control devices, sampling instruments, calibration tools, and other equipment and consumables that have potential impact on QEH&S. The inspection and evaluation process is intended to verify that materials are suitable for their intended purpose and in good working condition upon receipt. This review should also include an evaluation of safety information accompanying hazardous or toxic materials operator manuals for equipment.
- Does the organization retain appropriate QEH&S inspection and evaluation records that clearly document the inspection and evaluation? Does the documentation confirm that authorized personnel perform the inspections and that corrective measures are identified?
- Does the organization: appear to understand which processes require some type of routine or priority inspection or monitoring procedures? Is follow-up status of inherently hazardous processes or activities evaluated?
- Does the organization conduct inspections and evaluations of those workplaces on a frequency commensurate with potential risks?
- Are there procedures and records for workplace QEH&S monitoring where appropriate?

7M9 Does the facility monitor key QEH&S aspects of operations related to the delivery of a quality product or service. Does the process include periodic inspections and audits to determine conformance with regulatory requirements and audits of the QEH&S management system. Are the multiple root causes of process/procedural policy or regulatory nonconformances identified and are corrective/preventive actions developed and followed-up until effectively executed? Is allocation of resources to the audit process, including management time, materials, technical, and financial support and training, provided?

- Has the organization established and implemented procedures to monitor, review, and evaluate the applicability of QEH&S policies, laws, standards, and regulations?
- Is there evidence that the calibration procedures are followed?
- Is there evidence that workplace environmental monitoring is conducted and that records of the results are maintained in accordance with corporate policy regulations, including changes to existing requirements?

- Independent management audits of the QEH&S program should be conducted at least annually. Is there objective evidence or documentation of program audit conformance to applicable requirements as necessary?

7M10 The results of audits are documented and reviewed, including the identification and mitigation of multiple root causes of process failures. Process corrective actions are followed-up to ensure process improvement and conformance, including the processes for customer complaints and reports of nonconformance. Is there evidence that corrective/preventive actions are initiated, including control procedure changes or changes made in other facilities or work areas, and did the potential for similar nonconformances or accidents/incidents exist?

- Nonconformances highlighted by the audits are prioritized and monitored to ensure corrective implementation.
- Are audit findings, conclusions, and prioritized recommendations submitted to appropriate members of the organization, including senior facility management? Typical reports should identify the specific system or process deficiencies and the root causes of those deficiencies.
- Although not required, periodic (as often as possible) job or task observations-monitoring to ensure that personnel are conforming with critical job requirements should be established and maintained with appropriate feedback to stakeholders. Desired behaviors and correction of nonconforming activities should be reinforced.

7M12 Has the facility identified inspection and test status of all products relating to QEH&S to ensure performance in accordance with design or process specifications? Are inspection and testing of specified incoming items, services, and processes conducted by using these performance criteria?

- Is inspection and test equipment calibrated and maintained as required by this standard and other relevant specifications, including manufacturer requirements?

7M13 All sample collection, test and measurement equipment, or analytical devices must be maintained and calibrated in accordance with manufacturer or regulatory requirements and/or internal policy. Records of all calibration and maintenance procedures must be maintained and referenced or included in reports by using data from use the calibrated devices. Is calibration documentation verified before using test or sample collection equipment?

- Is inspection and test equipment calibrated and maintained as required by this standard and other relevant specifications, including manufacturer requirements?
- Have calibration criteria been defined and documented for inspection, measuring, and evaluation equipment at prescribed intervals, including

referencing the standard or criteria for such calibration? The organization should document SOPs for the use of sampling and analytical equipment.

- Are the measurement devices calibrated by a valid methodology and labeled? Check measurement device instruments and review the instrument's calibration label and calibration records. Review the calibration expiration date to determine if it has expired.
- Are calibration records maintained in sample files and in calibration logs for all test equipment?

7M14 Has the facility documented the defined criteria for selecting measuring and test equipment with regards to accuracy, precision, and environmental demands?

Documentation/Records

Audit plans, inspection, or job observations checklists

Audit reports, corrective actions, and follow-up documentation

DOCUMENTATION AND RECORDS

References: ISO 9001: 4.16, 4.5; ISO 14001: 4.3.4, 4.3.5, 4.4.3, 4.5.3

Records should be established for the following:

- Policy
- QEH&S Manual
- Process control instructions, SOPs
- Contracts
- Design
- Purchasing
- Product ID and tractability
- Inspection, testing, and measurement, including calibration
- Nonconformance, accident or incident investigations
- Corrective action or preventive action reports
- Training
- Audit reports and management review

7M15 The Q EH&S management system is described in a manual that references related documentation, the requirements for and maintenance of management system records, or records dictated by or associated with all related QEH&S processes or procedures. Procedures are established for the identification, collection, indexing, filing, storage, maintenance, and retrieval of all quality records with specific retention times defined for each type of record.

- Has the scope of the document control procedure been defined?

7M16 Procedures are periodically reviewed and modified as necessary to conform to process, facility, regulatory, customer, or industry standards or policy. Are processes in place for the review-approval and sign-off of all control procedures?

7M17 Current versions of all relevant documents and procedures are available and in locations convenient for personnel or stakeholder access to effectively conform to management system requirements. Does the document control procedure identify crucial external QEH&S documents including laws, standards, and regulations?

Note: Verify that the organization possesses the most recent version of applicable laws, standards, and regulations.

7M21 There is a process for removing obsolete documents and procedures (especially work process control instructions) and replacing them with current documents. The facility also maintains archives for specific records required by regulation such as QEH&S-related procedures. It is recommended that documents be identified with revision number and date in the footer and that there is a list or inventory of all control documents.

- Does the organization maintain a document control procedure that defines crucial internal QEH&S documents, including policies or procedures originating within the department, business unit, corporation, and/or parent organization?
- Is there evidence that existing documents adhere to the document control procedure and that appropriate resource allocation is committed to the process?

Note: Review select documents to assess adherence to the defined control procedures and the integrated quality and EH&S management system.

- Is there a master list of control procedures identifying the current revision status of the documents?
- Is there a system to enable the removal of outdated or obsolete EH&S documents?
- If superseded or obsolete control documents are stored. does the organization maintain a method whereby the old documents can be identified as obsolete?
- Does control of superseded or obsolete control documents include measures to ensure use is limited to current documents?

Note: Take samples of documents from the master list or equivalent list as well as applicable revisions; proceed to distribution location, then physically verify that the distributed site/department has the latest revision. Check to ensure that the document is authorized. The verification approach described above may also be conducted in reverse. For example, take samples from the work area documents and check them against the master list.

- Are changes, reviews, and approvals recorded on controlled QEH&S documents?

7M22 Control all documents related to QMS/EMS or other systems such as Responsible Care or QS 9000. Determine document retention times, and establish a master List of control documents and indexed system for records maintenance and retrieval, including the removal and replacement of obsolete documents (e.g., Mgt. system policy and manuals, EH&S regulations, audits, inspections and calibration records, job observations records, maintenance, medical, personnel and training records).

Is there a system for QEH&S document review and approval prior to issue? Are the reviews noted, including review/sign off by qualified persons, and stakeholders?

Note: There are many types of controlled documents. They may be procedural documents, planning documents, and product documents. Controlled documents share the following characteristics.

- Reviewed and approved by authorized personnel prior to issue
- Changes are reviewed and approved by authorized personnel prior to re-issue.
- A master list or other document control procedure is established to identify the current issue of the documents and the date of the changes to preclude the use of non-applicable documents.
- Controlled documents are identified with the date, version, and approval of the documents. There is no requirement that all documents be controlled. Changes to control document should also be documented on document control ERRATA sheet.

Documentation/Records

- Master list of control documents and distribution list, process
- Document approval, revision process, approval verification
- Process control, inspection, maintenance, and servicing records, including calibration
- Sample plans, measurement, and statistical procedures
- Inspection/testing records
- Personnel qualification records

- Inspection, job, or process observation checklists
- Inspection monitoring, testing, and calibration procedures

CORRECTIVE AND PREVENTIVE ACTIONS

References: ISO 9001: 4.14; ISO 14001: 4.3.7, 4.5.2

7M20 All accidents, product/service, or nonconformances must be evaluated to determine the multiple root causes of the process failure and to establish interim and long-term corrective/preventive actions to mitigate the risk.

- Do prioritized corrective actions result from inspection, nonconformance investigations, audits, and evaluation efforts, and are supervisors and employees informed of investigation or inspection results?

Note: Verify that prioritized corrective actions are generated and followed-up

- Do qualified individuals prioritize corrective/preventive actions based on the severity and probability or by another appropriate methods?
- Are effective interim control measures implemented until a state of compliance or conformance is achieved?
- Do documents verify that corrective actions have been taken?
- Are closure of corrective actions systematically reported to management?
- Are there documented procedures for implementing corrective/preventive QEH&S actions?

Note: This point refers to the general approach the organization uses to prevent quality or EH&S risks.

- Have the QEH&S corrective and preventive action procedures been distributed throughout the organization?
- Where revisions have been provided of corrective or preventive actions, have the changes been documented and communicated to appropriate individuals?

7M21 Have the responsibilities for investigating accidents and process nonconformance been established by policy and do they include work center supervisors as well as technical or operations engineers and experts to support the process? Is the process understood and does it facilitate the identification of multiple root causes of process failure?

Documentation/Records

Nonconforming material
Product or service reports

Customer complaints
Accidents–incident investigation reports, including corrective actions and follow-up

RISK AND OPERATIONAL CONTROL

References: ISO 9001: 4.15, 4.9, 4.19, 4.7, 4.13, 4.14; ISO 14001: 4.3.6

8M1 Verify, store, and protect customer-supplied tools, procedures, or products that will be used or incorporated in the delivery of their service or product. Customers must be notified in the event that these resources are lost or destroyed. Does the facility have a documented process for this requirement?

Are customer-supplied goods and services subject to risk assessment/inspections prior to use, and are records maintained to verify this procedure?

Documentation/Records

SOPs, production or service delivery plans
Maintenance plans/schedule
Personnel training qualification records

DESIGN, PROCESS AND ENGINEERING CONTROL

References: ISO 9001: 4.4, 4.8, 4.9, 4.10; ISO 14001: 4.3.6

8M2 Identify and document plan or process design reviews. Ensure that competent-experienced personnel perform design and procedures and, where appropriate, that the review is completed with tools such as systems safety, MORT or qualitative risk assessment. All design reviews must be documented and changes or corrective/preventive actions must be tracked to completion.

- Does the organization maintain documented procedures for assessment of nonconforming processes or devices, including periodic follow-up?

Note: Review established operations and maintenance (O&M) procedures. Do O&M procedures appear adequate for the nature and scale of processes and devices used by the organization?

- Do purchasing personnel have a current inventory of all hazardous materials, and is there purchasing guidance and technical concurrence for the procurement of hazardous material?
- Do work centers know how to get material safety data sheets (MSDS's)?

- Is there an inventory of high-risk operations, prioritized by degree of quality or QEH&S risk by shop, work center, or supervisor?

8M3 Does the facility verify product/service design to ensure completeness and that it meets required customer, regulatory, and process requirements? Has the organization ensured that technical interfaces between different input groups are established, including regulatory and/or the results of any corrective/preventive action reviews (lessons learned)?

- Do supply personnel have procedures and do they understand the process for HM review or the procurement of quality or EH&S controls?
- Is QEH&S considered in the context of equipment used or purchased for production, installation, and maintenance of facilities, tools or equipment?
- Is there a review/inspection process to identify unauthorized modifications of tools or equipment by operators attempting to facilitate production requirements?
- Are workplace designs and equipment layouts examined for potential EH&S deficiencies?
- Does the organization have procedures to identify QEH&S job performance requirements?

8M4 Is there a formal procedure for reviewing and giving signature approval to design or process changes? Design output must be documented against design input. Are reviews verified and documented at appropriate stages of the design process?

- Is there a process for modifying job procedures to comply with regulatory requirements and standards of practice?
- Is there evidence that work instructions and procedures are followed-up?

8M5 Document customer design product/service delivery or output requirements, including a formal system to verify changes and validate deliverables to ensure that they meet user needs.

8M6 Management plans production, installation, and servicing procedures and ensures that these processes are executed. Personnel are provided with procedural control instructions, equipment, material, and tools suitable to ensure product or service delivery. Procedures must also comply with applicable standards, codes, or regulations.

- Other required procedures:
  Contract review procedures
  Design procedures
  Document and data-control procedures

Purchasing procedures
Product identification and tractability
Process control procedures
Servicing and maintenance procedures
Inspection, testing, measurement and calibration procedures
Corrective and preventive action procedures
Handling, storage, shipping, disposal, and labeling of products, HM/HW
Internal audit procedures
Training
Statistical procedures

- Is there evidence that individuals conducting the assessments are trained in hazard identification and risk assessment techniques and critical job or task observations?
- Is there evidence that assessments are undertaken by individuals with specialist expertise (e.g., ergonomists, hygienists, safety engineer, etc.)?
- Is there a risk assessment on potential for on-site process and off-site emergencies?
- Are examples of health and safety risks being considered prior to the decision to purchase equipment, contract services, or purchased materials?

8M7 Has the facility developed documented procedures consistent with facility, ISO, industry and regulatory standards? Organization/Facility policies address integrated process improvement, and quality and EH&S requirements.

8M8 Has the facility established and executed process/procedures for testing to ensure that products/services meet customer and internal quality standards? Performance or workmanship requirements must be clearly specified and documented. Processes must be carried out by qualified personnel and/or require continuous monitoring and process control to ensure that requirements are met and that work is performed safely and without creating QEH&S risk or liability.

- Do employees have access to necessary documents and/or information at locations where essential operations are performed? Are process changes or lessons learned from process failure multiple root cause analysis or accident investigations communicated to stakeholders?
- Is this documentation available in a format appropriate to its use and to the training of the employees at that location?
- Are invalid and/or obsolete quality, process, and EH&S documents removed or their use otherwise prevented?
- Does the organization maintain a procedure for identifying and retaining obsolete EH&S documents that may have future relevance? Do employees have access to appropriate documents and/or information at locations where essential operations are performed, including process change communication

or lessons learned from process failure multiple root cause analysis or accident investigations?

- Is this documentation available in a format appropriate to its use and to the training of the employees at that location?
- Do potentially hazardous operations have adequate QEH&S procedures?
- Are safety and environmental control requirements specified in SOPs or process instructions?
- Are personnel informed of the procedure, for example, process briefings and/or inclusion in training?
- Do documented administrative and technical procedures define the manner of production, installation, and servicing of tools and equipment?
- Has the organization established and documented safe work practices (or equivalent procedures) into existing procedural controls for those activities that could cause serious physical harm, environmental liability, or process quality/rework issues?
- Does the organization implement procedures to ensure QEH&S performance maintenance activities?
- An effective QEH&S system includes proactive maintenance and servicing of production equipment. The organization should maintain a comprehensive safety and maintenance program to maximize equipment reliability and prevent breakdown of equipment, injuries, and environmental liabilities?
- Is there a lock-out/tag-out process that ensures plant or tools are removed from service until competently repaired.

Note: Unsafe or out-of-service equipment must be easily identifiable so that it is not used by mistake. This arrangement may be done by placing an identifying mark or tag on it or by isolating it in a designated area. The system must be documented and communicated to appropriate employees.

- Are there controlled access procedures, including signs and systems to delineate restricted access areas and to prevent access to controlled areas?
- Is there evidence that controls are checked, for example, inspection checklist reports, minutes of health and safety committee meetings, or maintenance?
- Do supervisors ensure that tasks are performed safely and in accordance with work instructions and procedures?
- A maintenance request procedure includes provision for unsafe tools or an equipment plant to be withdrawn form service (lock-out/tag-out). Where necessary, there is a lock-out system to prevent the untimely start-up of plant.
- Are there records documenting details of inspections, maintenance, repair and alteration of tools, or equipment?
- Do plants that require registration have a current certificate of registration or permits?

- Are there documented permit to work procedures for high-risk or critical processes, plants, equipment, and materials?

Documentation/Records

Program plans, schedules
Design and design review documentation
Standards, specifications for products, services and procedures
SOPs, work instructions, change orders

TRAINING

References: ISO 9001: 4.18; ISO 14001: 4.3.2

8M8 Is training necessary for the implementation or maintenance of the QEH&S management system, or any related procedures or processes effectively executed and documented?

- Is QEH&S training included as part of overall job training?
- Does the organization provide employee and contractor quality and EH&S training to promote the knowledge and skills required to successfully fulfill job performance requirements, including safety and the identification and control of process hazards/risks?
- Has the organization identified and documented the minimum skills, education, and attributes of QEH&S training personnel?
- Do personnel receive QEH&S refresher training as required by policy or regulatory requirements?
- Can workers demonstrate comprehension and retention of training, verified by focused job or task observations or behavioral-based safety observations?
- Does the organization have procedures to identify QEH&S job performance requirements?
- Has the organization developed QEH&S training educational objectives or equivalent benchmarks?
- Has the organization developed a documented procedure to ensure that personnel achieve the objectives of QEH&S training or documented/reinforced desired behaviors and process conformance through job observation and performance measurement? Has the organization established documented procedures for identifying QEH&S risks through the various stages of purchasing, production, delivery, and maintenance?
- Have these procedures been established and maintained by qualified personnel?

8M9 All personnel (management, staff, or contractors) who can potentially impact the quality or EH&S performance must be trained. The training objective is to understand the risk associated with the activities and the procedures that must be followed to mitigate the risk. Training must also include emergency preparedness and response procedures.

- Are tasks allocated to staff based on ability and documented training?
- Is there evidence that competency is established through a skills or training register?
- Is there a maintenance procedure that includes a reporting mechanism and action for removal of unsafe equipment from service, including a lock-out/tag-out system to prevent the use of equipment and to isolate and prevent machinery from becoming operational (e.g., during maintenance)?

8M10 All personnel are trained to understand the QEH&S policies for the facility, including their respective roles, to comply with the policy and achieve its defined objectives and milestones. The training shall, at a minimum, make employees aware of potential process QEH&S risks and hazardous conditions, including symptoms of overexposure to hazardous materials and the methods to reduce or control the risk of exposure as well as their roles/responsibility to the system.

- Do job descriptions identify specific requirements relevant to QEH&S?
- Is required/necessary training documented?
- Does the organization verify that QEH&S training adequately provides personnel with the knowledge and skills to understand emergency control of equipment and procedures?
- Verify that the organization assigns qualified process QEH&S personnel to critical job tasks.
- Has a documented training needs analysis has been conducted that included QEH&S requirements and the preparation of individual development plans (IDPs)?
- Does training consider different levels of ability and literacy?
- Do trainers have appropriate skills and experience?
- Are facilities and resources suitable for effective training?
- Does the organization document and keep records of all training?
- Is every training session evaluated to ensure comprehension and retention?
- Is the training program reviewed regularly to ensure its relevance and effectiveness?
- Is training is provided to affected personnel when there are changes or processes in the workplace?

8M11 Are personnel assigned to perform tasks that could impact the quality of the product or service outcome or the EH&S associated with assigned tasks competent to perform their duties based on education, training, or experience?

- Have members of executive and senior management participated in training that includes an explanation of legal obligations, QEH&S management principles, and practices?

PURCHASING

References: ISO 9001: 4.3, 4.6, 4.8, 4.12; ISO 14001: 4.3.6

8M12 Does the facility maintain documented procedures for contract review to ensure that requirements are adequately defined and that the customer has agreed to them before the commencement of services or product design/delivery? In addition are there procedures for amending a task order or making contract changes?

- Does the operations manager coordinate the review of contracts or contract changes, including required materials or equipment before assigning resources or purchasing the materials?
- Is there a documented purchasing procedure that outlines how QEH&S is considered prior to purchase of goods or services or the use of customer or contractor-supplied tools and materials (which may be incorporated into a quality procedure)?

Verification: Review contracts and orders to ensure customer requirements are adequately defined and that any differences are mutually resolved.

HANDLING, STORAGE AND PACKAGING OF MATERIALS, INCLUDING HM/HW

References: ISO 9001: 4.6, 4.8, 4.10, 4.15; ISO 14001: 4.3.6

8M14 Is the process for purchased products, services, and subcontractors controlled to ensure delivered products and services meet requirements?

8M15 Are product receipts inspected to verify if the material does not meet design, contract, or purchase specifications and to trace products at all stages from receipt of raw materials to installation, use, and disposal?

8M16 Has the organization established procedures to handle, transport, ship, or store materials or product to prevent damage or the accidental release of hazardous material, including product packaging requirements? How does the facility ensure the quality and safety of products after manufacturer or receipt before distributing to customer or stakeholder?

- Are on-site activities that involve the use, generation, storage, treatment, and disposal of HM/HW controlled, including the storage, segregation, and transportation?
- Review hazardous waste manifests to verify types of wastes generated.
- Is there documentation of HM/HW releases at the site? What was the quantity released, location of release, description of remedial actions taken, description of known or suspected impacts on the physical environment, and regulatory reporting, if applicable?
- Is the site a generator, treater, storer, disposer, or transporter of HW?
- Is the facility in compliance with HW regulations?
- Does each site or work area have information on the HM used in this area:

  MSDS's

  Containers labeled

  Written program

  Employee training
- Is there an inventory and site map identifying all HM/HW storage areas?
- Has the company implemented a employee and community right-to-know program?
- Has this information been incorporated into emergency response plan documents, including external communications with the fire department?

Does the HM/HW business plan include the following?

Faculty identification

Owner/operator name

Phone number

Address

SIC code

Dun & Bradstreet number

Emergency contact/notification

Inventory information

Site map

- Are all storage areas protected and contained to prevent/minimize releases?
- Does the site comply with all labeling and posting requirements for containers and storage areas, including HW?
- Does the facility have procedures for mitigating HM/HW releases, including supplies, equipment, and employee training?
- Are reporting thresholds identified and reporting methods explained?
- Are incident critique and follow-up procedures in place?

- Are all hazardous wastes stored or accumulated in secondary containers?
- Are incompatible HM/HW segregated?
- Are all containers of HW inspected weekly, and is documentation maintained, including dated labels?
- Are all ignitable or reactive wastes kept at least 50 feet from the facility's buildings and other flammable or combustible materials?
- Are waste manifests maintained for at least three years from receipt at the waste-designated facilities?
- Are there discharges of waste-to-storm drains or surface-, or groundwater? If so, what is discharged and does the site have permits? Is the best conventional pollutant technology (BCT) used prior to discharge?
- Are potential employee exposures to HM/HW documented, and are they informed of monitoring results?
- Are there noise-hazardous operations?
- Have potential high-hazard areas or processes been identified?
- Are chemical tanks and piping systems labeled?
- Do chemical feed systems have measuring devices to record volumes of material used or released?
- Is there an inventory of all control ventilation systems?
- Are these systems evaluated annually, including preventative maintenance?
- Are hoods labeled with desired velocity or static pressures?
- Is there a documented program for respirator and personal protective equipment (PPE)?
- Is there a list of operations that require PPE ?
- How are respirators and PPE cleaned or replaced?
- How are respirators fitted, and what records are maintained for training and fitting of this equipment?
- Do supervisors enforce the wearing of PPE constantly?
- Is HW stored in excess of 90 days?

Do all storage requirements meet the following standards?:

Segregation
Berms
Fire extinguishers
Emergency spill kits
PPE

8M18 Do the facilities routinely inspect the quality and usability of products in storage or materials to be used in the production process?

Verification: Observe chemical storage for appropriate containment, fire protection, and compatibility of storage.

- Are HM/HW and damaged or nonconforming products segregated and clearly labeled or tagged?
- Have facilities established requirements for material storage, spill containment, labeling and transportation (e.g., rack height, ventilation, containment, fire protections, grounding, or bonding?
- Are there procedures that ensure that materials are disposed of in accordance with relevant regulatory requirements?
- Are there copies of MSDS's, with review dates, held in work areas or are they readily accessible electronically?
- Is there evidence that MSDS's are checked for applicability, currency, legibility, and availability?

Note: MSDSs provide important information about hazardous substances, including health hazard information, precautions for use, and safe handling information. MSDS's should be readily available and accessible to personnel who handle HM/HW.

- Are there are systems for identifying and labeling HM/HW?
- Is there a procedure that ensures HM/HW are appropriately identified and labeled at all stages in the production process?
- Are the smaller containers in which materials are decanted clearly labeled to identify the contents?
- Where materials are produced, procedures ensure that all materials are identified and clearly labeled in accordance with regulatory requirements, including the development of MSDS's?
- Are employees familiar with HM/HW management controls and laws?
- Is there a documented purchasing procedure that outlines how QEH&S is considered prior to purchase of goods or services or the use of customer or contractor supplied tools and materials (which may be incorporated into a quality procedure)?
- Is a system or listing of relevant health and safety related information accessible by those recommending a purchase?
- Are examples noted of where QEH&S risks have been considered prior to the decision to purchase?
- Has the organization systematically identified and inventoried all HM/HW under its control?

8M19 How does the facility identify, tag, or label all unique products, batches, or tools required for contract or product outcome to ensure traceability?

8M20 How does the facility ensure that all conforming products are packaged/preserved and shipped in such a way that its protection and product conformance is ensured when it arrives at its destination.

Documentation/Records

Special handling procedures

Shipping, treatment, disposal, and bringing response procedures

Inspection and corrective action reports

MANAGEMENT OF CONTRACTORS AND SUPPLIERS

References: ISO 9001: 4.3, 4.8, 4.12, 4.14; ISO 14001: 4.3.6, 4.42

8M21 Is there a vendor qualification list and how were the vendors qualified?

8M22 How does the organization evaluate subcontractors' credentials, including verification of prior performance on contracts and establishing a system for on-going assessment of their work during the contract period?

- How does the facility confirm the capacity of the vendor or subcontractor to fill the order or provide service in accordance with specifications on time and within budget?
- Does the organization have a systematic contractor selection and evaluation method?
- Does the organization document contractor quality and EH&S performance evaluation and assessment?
- Is the selection of contractors contingent upon the offer's ability to meet specified quality and EH&S requirements?

8M25 Does the facility precisely describe contract assumptions, goals, objectives, and anticipated schedules, including a description of the product or services? Is signature approval on the purchasing documents necessary prior to being issued to the vendor or subcontractors?

- Does the organization document vendor or subcontractor quality and EH&S performance evaluation and assessment?
- Is the selection of contractors contingent upon the offer's ability to meet specified quality and EH&S requirements?

Documentation/Records

Approved vendor, subcontractor list

Procedures/checklist for purchase, vendor, subcontractor review

Purchase order and contracts

Subcontractor qualification records, including training

Contractor, subcontractor performance monitoring
Purchase contract review procedures

EMERGENCY-CONTINGENCY PREPARATION

References: ISO 9001: 4.14; ISO 14001: 4.3.7

8M26 Has the facility identified and evaluated the potential for natural disasters and facility or transportation risks? Have all identified emergency conditions been well- defined, including response plans drills and community emergency response support procedures as appropriate?

8M27 Are all drills or emergency response actions critiqued by senior managers and emergency response personnel to identify the need for process improvement or to correct observed nonconformance or inadequate/partial response to response procedures?

- Is there an emergency plan or a document for those in charge of emergency response?
- Is emergency response training documented?
- Is there a method to assist employees to identify responsible person(s) in an emergency situation (e.g. use of armbands or colored helmets to identify fire wardens)?
- Has the facility identified and documented potential emergency situations (both on-site and off-site)?
- Are there specific procedures for each type of emergency situations?
- Emergency instructions and emergency contacts are prominently displayed and made throughout the organization.
- Do the instructions include phone numbers and contacts, including after-hours contacts where appropriate?
- Have employees received instruction in emergency procedures appropriate to the degree of risk, including first aid and CPR?
- Is evidence of emergency procedures included in the content of induction or other training programs?
- Have potential emergency situations (both on-site and off-site) been identified and have emergency procedures been documented?
- Is there a risk assessment that assesses the potential for on-site process and off-site emergencies?
- Is there a formal system in place to regularly review emergency plans and instructions, including drills?
- Are competent persons determined and are emergency equipment and alarm systems inspected and tested? Also, has the suitability of location and accessibility of emergency equipment been determined?

MAINTENANCE/SERVICING

References: ISO 9001; ISO 14001

NONCONFORMANCE AND ACCIDENT INVESTIGATION

References: ISO 9001: 4.13, 4.14; ISO 14001: 4.4.2

9M1 There must be a system to determine the multiple root causes of customer complaints, accidents, and product or process nonconformances (including near-misses that could have resulted in a serious incident). Changes in the process design as a result of lessons learned should be executed to correct/prevent a reoccurrence of the accident or nonconformance.,Are staff and management (all levels) trained to perform multiple root cause investigations?

- Are investigations also included in emergency plans and procedures?
- Is multiple root cause identification an objective of QEH&S nonconformance investigations?

Note: Review representative nonconformance or accident reports or equivalent documents to that verify multiple root cause analysis is systematically conducted.

- Are employees trained regarding a documented system to ensure that all workplace injuries, illness, and incidents are reported? Do their supervisors provide this training? If yes, how is this training documented?
- Is there evidence that procedural noncompliance is detected and documented and that corrective/preventive actions are executed? Monitoring or measurement should include positive reinforcement of process conformance.
- How does the facility comply with HM/HW regulations, including transportation rules for shipping, procurement, minimization, storage, transportation, or disposal of HM?
- Are conclusions and lessons learned from such investigations promptly communicated to management and employees potentially affected by the process changes, including similar facilities or on-site activities?
- Do investigation reports contain recommendations and dates for implementation of corrective actions, and how is implementation/follow-up documented?
- Is responsibility for and target completion dates assigned for implementing corrective actions?
- Has senior management concurrence for corrective/preventive actions been obtained prior to issuing corrective action reports or implementation of process changes?
- Has the effectiveness of corrective/preventive actions been evaluated and confirmed?

- Does the organization maintain procedures to document investigations, work-related fatalities, injuries, and illness or process failures, including near-misses (incidents) or releases of HM/HW? Note: organizational policy or applicable regulations may dictate the scope and content of both internal and external reporting.
- Does the management review and provide signature approval for work-related fatality, injury, illness reports, identification of multiple root causes, and recommended corrective actions?
- How are lessons learned shared with other supervisors and employees (e.g., briefings, correcting procedures, etc.)?
- Who initiates an investigation and are direct and indirect costs estimated?
- Is there a manager responsible for medical case management? Please describe how the system works and provide the name of the person to contact.
- Is systems safety, risk assessments, or job-task analysis used to identify and evaluate high-risk job processes? What is the system/process design review and does it include professional quality, safety, environmental review, and sign-off?
- How are reviews evaluated and corrective actions implemented?
- How are recommendations executed and followed-up to ensure completion?
- Obtain and review a copy of the company's accident investigation form and policies, training, etc.
- Does senior management review accident and nonconformance reports and ensure that *all* corrective actions are complete?
- Is the major objective of accident, incident, and illness investigations the identification and correction of the multiple root causes of process failures, accidents, or incidents?

Note: An effective illness/incident/accident investigation should, regardless of length or style, document applicable facts. Conclusions should be reached only after interviews with all pertinent individuals have been completed. Six key questions should be answered in the investigations: who, what, when, how, and, most importantly, root cause determination (i.e., why). In addition, investigators should document, to the extent possible, direct and indirect costs for the nonconformance or accident.

MANAGEMENT REVIEW AND CONTINUOUS IMPROVEMENT

References: ISO 9001: 4.2; ISO 14001: 4.5

An organization should regularly review and continually improve its EH&S management system with the objective of continuous improvement of its QEH&S performance and its compliance with relevant standards or guidelines.

10M1 Does senior management periodically review the QEH&S management system to ensure that the process for continuous improvement is maintained and effective and that it meets operational and regulatory objectives?

10M2 There must be a defined process for the risk manager, quality manager, and EH&S manager to prepare integrated reports with the results of audits and identified nonconformances, including their multiple root causes and a plan to correct/prevent the process failure or management system nonconformance. Other management system progress reports include cross-facility, consultant, registrar audit reports and recommendations, changes in regulations, ISO, or industry standards.

- Is trend analysis periodically performed on findings to identify underlying programmatic or management root causes?

Note: senior management should periodically review all aspects of the organizational QEH&S performance to determine if similar deficiencies exist across different departments or activities. Management assessments should focus on the identification and resolution of systemic and cultural management issues and the multiple root causes of process nonconformance. Strengths and weaknesses affecting the achievement of QEH&S goals and objectives should be identified so action can be taken to improve quality across the entire organization.

10M3 Does the management review document considerations regarding process changes, changes to policy, and management system objectives and milestones, as well as well as process changes, changes to policy and management system objectives, and milestones?

OFFICE QUALITY ASSESSMENT NONCONFORMANCE REPORT TRACKING FORM

Facility Location: ______________________________

Date: ______________________________

Reference Section #: ____________________ Nonconformance #:

____

Audit Question:

Findings/Observations and Comments:

Recommendation(s):

Responsible Manager:

Auditor and Auditee Use Only:

Target Corrective Action Date: ______________________________

Assigned to: ______________________________

Assigned by: ______________________________

Follow-UP/Verification Findings:

Verification of Corrective Action

Confirmed By: ______________________________

Date: ______________________________

Appendix II

Miscellaneous Environmental Health and Safety Program Audit Attributes and Technical Check Sheets

MEDICAL AND OCCUPATIONAL HEALTH

1. Describe how workplace injuries are treated. Are the facilities designated urgent for medical care within 30 minutes or 15 miles of the site? Does management consider this arrangement satisfactory?
2. Does the facility have written policies and procedures for managing medical emergencies?
3. Is there a current documented assessment of workplace first-aid requirements (e.g., number of first-aid administrators, level of first-aid training, number of first-aid kits, first-aid equipment, and special requirements, such as antidotes, emergency evacuation, and treatment?
4. Is there evidence of inspections of first-aid facilities, kits, and equipment?
5. Are there records of first-aid treatment?
6. Are medical pre-placement and periodic examinations provided for all employees required to wear respirators, or is there work in areas where potential exposures to HM/HW exceed the action level, including records of health monitoring? Is the health monitoring performed in accordance with regulatory or industry best practices?
7. Is there at least one employee on each shift currently qualified to render first-aid?

8. Is there a return-to-work policy?
9. Does the company utilize physical or occupational rehabilitation specialists for disabled employees collecting compensation?
10. What percentages of claims are due to injured backs or repetitive motion?
11. What employee groups, departments, or work centers have the highest frequency of injuries? Is the hospital or medical clinic familiar with or prepared to treat workplace injuries and/or emergencies?
12. Describe the company's back injury prevention or ergonomics program?
13. Does the company have an occupational health nurse, safety engineer/coordinator or industrial hygienist on staff?
14. Are personal hygiene requirements enforced for employees so that they do not carry hazardous materials or waste home with them or experience secondary exposures when they handle food or cigarettes?
15. Is there a documented procedure and schedule for monitoring screening and testing?
16. Are confidentiality provisions included in the procedures, and are these provisions followed?
17. Is health screening required for job applicants or for employees assigned to a new job? Does this include a standard medical history and physical examination?
18. Are job-specific examinations performed by a physician experienced in occupational medicine and knowledgeable of your company's special requirements (job demands)?
19. Are your company management and supervisors trained regarding the employee assistance program and how to recognize employees that may need assistance?
20. Are employees encouraged to use the employee assistance program (such as newsletter, electronic mail...)?
21. Is there a qualified employee/consultant assistance program coordinator to oversee the program, (which include alcohol and drug dependence, physical health, and childcare...)?
22. Is there an in-house or an off-site fitness center for all employees?
23. Are health education and screening seminars (smoke cessation, nutrition and weight loss, stress control, mammography, cholesterol analysis, etc.) offered to all employees?
24. What employee groups, departments or work centers have the highest frequency of reported injuries?

ELECTRICAL SAFETY

1. Are electrical devices and electrical cords grounded, and are ground fault devices used as appropriate? Is the grounding periodically tested?

2. Are ground-fault circuit interrupters installed on each temporary 15 or 20 ampere, 120 volt AC circuit at locations where construction, demolition, modifications, alterations, or excavations are being performed?
3. Are circuit breakers, fuses, sized to fit the equipment in use?
4. Are electrical wires protected form chaffing, pinching? Are they free of splices, cuts, or breaks in their insulated coats? Are the electrical cords kept off of the floor or protected from vehicle and personnel traffic?
5. Are battery compartments vented to prevent the accumulation of hydrogen gas?
6. Are materials and equipment that can generate static electricity grounded?
7. Are electrical workers trained in CPR?
8. When working on electrical equipment, do workers stand or rubber mats?
9. Are all uninsulated conductors such as bus bars, panel boards, or switchboards enclosed in protected areas?
10. Are electrical panels accessible and unobstructed (panels must have a minimum clearance 36" deep and 30" wide), and are all unused openings plugged and is the cabinet fitted with a tight sealing door?
11. Are high-voltage control panels closed and secured?
12. Do switches show evidence of overheating?
13. Are all switches marked to show their use?
14. Are portable lamps guarded?
15. Are electric motors clean and free of excessive grease, and are the bearings in good condition to prevent overheating?
16. Are only authorized personnel allowed to work on or near exposed electrical sources?
17. When electrical equipment or lines are serviced, maintained, or adjusted, are necessary switches opened, locked-out, and tagged?
18. Are portable electrical tools and equipment grounded or double insulated?
19. Does the facility have electrical installations in hazardous dust or vapor areas? If so, do they meet NEC or CE requirements for hazardous locations?
20. Is the location of electrical power lines and cables (overhead, underground, underfloor, other side of walls, etc.) determined before digging, drilling, or similar work is begun?
21. Are all energized parts of electrical circuits and equipment guarded against accidental contact by approved cabinets or enclosures?
22. Are employees prohibited from working alone on energized lines or equipment?
23. Is stored energy released prior to performing work on equipment, for example, capacitors?
24. Is explosion-proof equipment used where flammable gases, vapors, dust, or other easily ignitable materials are present?

25. Are fenced or other secured barriers in place to prevent access to high-voltage transformers and the like?
26. Are the non-current-carrying metal parts of electrically operated machines bonded and grounded?

PERMIT SYSTEM: HOT WORK AND CONFINED SPACE ENTRY

The Permit

1. Do permits specify the job to be done and to whom the permit is issued, including its expiration date and time?
2. Does the system require that the potential hazards at the work site to be identified and recorded on the permit, including the precautions to be taken by the employee and the issuing authorities? Do permits specify the area to which work must be limited?
3. Do the permit recipients sign the permit to show that they have both read it and understood the conditions required for its use/approval? Are there provisions on the permit form to cross-reference other relevant certificates and permits?
4. Is there a requirement that work being done under a permit be stopped if any new risks are recognized?
5. Does the permit contain rules about how the job should be controlled or abandoned in the event of an emergency?
6. Where there are isolations common to more than one permit, is there a procedure to prevent the isolation from being removed before all the permits have been signed-off?
7. Are the types of work, types of jobs, or areas where permits must be used clearly defined and understood by stakeholders, including those who may use a permit and who may authorize permits?
8. Does the permit system extend to contractors and their employees?
9. Is the permit system flexibly applied to potential risks that were not identified when the system was established?
10. Are individuals prevented from issuing permits to themselves? Does the permit system conform with the legal requirements applying to the site?
11. Does the lock-out program include a permit form? Have all permit-required confined spaces been identified?
12. Have all permit-required confined spaces been posted? Is there an inventory to identify the locations of all permit-required confined spaces?
13. Do supervisors periodically spot-check permit work areas to ensure that requirements are followed?
14. Is emergency equipment such as portable fire extinguishers, self-contained breathing apparatus, protective clothing provided, and are they identified on the permit?

15. Do the permit forms such as lock-out, confined space entry, hot/cold work, hazardous material disposal, and rigging contain the minimum information required by the applicable standards, including authorized work dates?
16. Have supervisors and employees received documented training for permit programs requirements?
17. Is there a periodic audit of completed permit forms?
18. Does the permit for confined space entry include the hazards of the space, measures to isolate, and requirements or procedures to enter the space and does it eliminate or control hazards before entry and acceptable entry conditions? What about pre-entry and periodic atmospheric test, names of testers, and the times when tests were performed?
19. Is a buddy system with communications and an emergency retrieval harness/lanyard and lifting device required when entering tanks?
20. Does the permit for confined space indicate additional permits, such as hot work, that have been issued to authorize entry in the space?

Hot Work

1. Does the hot work permit require assigned fire watchers (individuals that are located at the welding site to identify and report emergency situations) whenever welding is performed where a fire could be started?
2. Are portable fire extinguishers and other fire protection equipment provided at the welding site?
3. Do hot work permits include requirements for welding components (such as steel members, pipes, etc.) that could transmit heat to combustible areas?
4. Do welders wear eye and face protection and fire-retardant coveralls and gloves?
5. When welding inside is performed, is ventilation positioned to capture metal fumes and other gases generated by the welding process?
6. Has the supervisor ensured that combustibles in the work area are protected from ignition prior to welding by moving them at least 35 feet and/or shielding them?
7. Is the hot work area swept clean, and are combustible floors, cracks, holes, and pipe openings covered with noncombustible material such as wet sand?
8. Before hot work is performed on metal partitions, walls, or pipes that pass through walls, are these areas checked from the opposite side to ensure there is no contact with combustibles?
9. Is cutting, hot work or weld drums, tanks, or equipment that contains or has contained flammable liquids or gases prohibited unless the containers have been cleaned, purged, and tested for the flammable vapors?
10. Is the work area monitored for up to four hours after work is finished?

Confined Space Entry

1. Are gases or combustible liquids used and stored in such a way that there is potential for them to leak into a confined space, forming an explosive, toxic, or oxygen-deficient atmosphere?
2. Are employees required to work in tanks or large sumps that could be oxygen-deficient or have flammable or combustible gases or vapors? Are appropriate engineering controls (ventilation) and administrative and procedural controls in place to test and control these environments to the extent possible before allowing personnel entry?
3. Do employees entering these types of spaces have adequate equipment (harnesses, personnel protective equipment, etc.)? Is entry allowed with a permit system, and a buddy system, which includes communications with the person entering the tank and others who may need to notified in the event of an emergency? Have employees been trained and have appropriate permits been completed for the entry and work in these control spaces?
4. Are confined spaces thoroughly flushed of any corrosive or hazardous substances, such as acids or caustics, before entry?
5. Are all lines to a confined space containing inert, toxic, flammable, or corrosive materials valved-off and blanked or disconnected and separated before entry?
6. Are impellers, agitators, or other moving equipment inside confined spaces locked-out before entry?
7. Is ventilation provided prior to confined space entry?
8. Is lighting provided for the work to be performed in the confined space?
9. Is the atmosphere inside the confined space frequently tested or continuously monitored during or work?
10. Are standby employees or other employees prohibited from entering the confined space?
11. Are welders and other workers nearby provided with flash shields during welding operations?
12. Is all portable electrical equipment that is used inside confined spaces either grounded and insulated or equipped with ground-fault protection?
13. Before gas welding or burning is started in a confined space, are hoses checked for leaks? Are compressed gas bottles forbidden inside confined space? Are only torches ignited outside of the confined area, and is the confined area tested for an explosive atmosphere each time before a lighted torch is to be taken into the space?
14. If employees use oxygen-consuming equipment such as salamanders, torches, or furnaces in a confined space, is sufficient air provided to ensure combustion without reducing the oxygen concentration of the atmosphere below 19.5 percent by volume?

15. Whenever combustion-type equipment is used in a confined space, are provisions made to ensure that the exhaust gases are vented outside of the enclosure?
16. Is each confined space checked for decaying vegetation or animal matter that may produce methane?

LOCK-OUT/TAG-OUT PROCEDURES

1. Have all employees working around equipment or energy sources that could be locked-out been trained to understand the requirements of permit and lock-out/ tag-out procedures?
2. Have lock-out/tag-out procedures been developed and documented?
3. Are the procedures actually used? Do procedures include the following?

   A specific statement of the intended use of the procedure?

   Specific procedural steps for shutting down, isolating, blocking, and securing machines?

   Specific procedural steps for the placement, removal, and transfer of lock-out/tag-out devices?

   Specific requirements for testing to ensure the effectiveness of energy control measures?

   Standardize within the facility in color, shape, size, print, and format?
4. Does management ensure that lock-out/tag-out devices are not used for any other purpose than the control of energy?
5. Is all machinery or equipment capable of activation required to be de-energized or disengaged, and blocked or locked-out during cleaning, servicing, adjusting, or setting-up operations?
6. Is lock-out of control circuits in lieu of lock-out of main power disconnects prohibited?
7. Are all equipment control valve handles designed for lock-out?
8. Does the lock-out procedure require that stored energy (mechanical, hydraulic, air, etc.) be released or blocked before equipment is locked-out for repairs?
9. Are devices attached at the same location, as a lock-out device would be?
10. Is new equipment or equipment that is replaced or undergoing major repair, renovation, or modification equipped with energy-isolating devices designed to accept a lock-out device
11. Are tag-out devices substantial enough to prevent inadvertent or accidental removal?
12. Do lock-out/tag-out devices identify the specific employee applying the device? Do tag-out devices:

Warn against hazardous condition?

Include a legend specific to the hazard (such as "do not start" or "do not open")?

13. Are audits of the energy-control procedures conducted at least annually?
14. Are audits performed by an authorized second- or third-party who does not have direct responsibility for use of the procedures being audited?
15. Prior to stating work on locked-out or tagged-out equipment, does the authorized employee verify that isolation and de-energization has been accomplished?
16. Are rotating parts of equipment, nip points, chains, gears, pulleys, and belts guarded to prevent the employee from accidentally coming in contact with the hazard?
17. Are operating controls locked when not in use to prevent inadvertent activation?
18. Are there emergency "Stop" controls positioned correctly, large enough, labeled, and color-coded to ensure easy identification and activation in an emergency?

RELEASE FROM LOCK-OUT OR TAG-OUT

1. Before releasing lock-out/tag-out, is there an inspection of the work area to ensure that nonessential items have been removed and that the machine is operationally intact, and all employees have been safely positioned?
2. Are affected employees notified that the lock-out or tag-out devices have been removed before starting of the machine or equipment?
3. Does the employee who applied the device remove lock-out and tag-out devices?
4. Are outside personnel informed of lock-out/tag-out procedures before engaging in maintenance and servicing operations?
5. Is primary responsibility given to one authorized employee during group lockout/tag-out efforts?
6. Do provisions exist for this authorized employee to determine the lock-out/tag-out status of individual group members?
7. Is primary responsibility given to one authorized employee during lock-out/tag-out efforts involving two or more groups of employees?
8. Does each employee of a group affix a personal lock-out or tag-out device to a group lock-out device before beginning maintenance or servicing? Does each employee of a group remove their personal lockout or tag-out device from the group lockout device when their work on the machine or equipment is completed?
9. Do procedures exist to ensure the continuity of lock-out or tag-out protection during shift or personnel changes?

10. Is there a maintenance procedure that includes reporting and removal of unsafe equipment from service?
11. Is there a tag system to identify tools or equipment that are in unsafe or out-of-service?

LIFE–FIRE SAFETY

1. Are manual pull stations and fire alarm evacuation devices tested semi-annually?
2. Do supervisors or other designated personnel periodically inspect work and storage areas for the presence and appropriate storage, containment, or disposal of combustible materials? Does this monitoring include weekly inspections of sprinkler water-control valves to ensure that they are locked open?
3. Are fuels stored separately from oxidizers and sources of ignition?
4. Are hot surfaces or others sources of ignition insulated to prevent contact with combustible liquids? Are overflow or other discharge drains protected from ignition sources?
5. Are fire protection, isolation, alarm, and separation systems in place and periodically inspected and maintained?
6. Are the building's public address (PA) system and other emergency communication equipment tested monthly?
7. Are smoking and "hot work" prevented in areas containing flammable or combustible materials?
8. Is there a sufficient number of exits for the occupancy loading of the facility, and exits are not obstructed in any way? Are exits marked and lighted with back-up power to ensure visibility if there is a loss of electrical power? Are these systems checked at least annually?
9. Are the correct class portable fire extinguishers for the potential fire risk mounted and readily available in each work area?
10. Are fire extinguishers inspected/tested monthly with the inspection documented on the extinguisher's inspection tag?
11. Are sprinkler heads protected and unobstructed by the material stored in the area? (There should be at least 18 inches' clearance)
12. Are interior stand pipes and valves regularly inspected?
13. Has the written emergency evacuation procedure been reviewed within the past 12 months?
14. Are current evacuation maps prominently posted on each floor?
15. Does the evacuation map indicate at least two emergency exits from the area?
16. Is the evacuation map properly oriented and the "you are here" site properly identified?

17. Are newly assigned employees notified of evacuation procedures?
18. Have employees been trained in emergency fire procedures and use of fire extinguishers, including periodic drills?
19. Have outside fire hydrants been flushed within the past year, and are they on a regular maintenance schedule?
20. Does the local fire department have a copy of the facility's emergency response and hazardous materials inventory-storage control plans for use in a fire or other emergency that may require going into or near these facilities?
21. Is the emergency generator exhaust directed away from the building and from air intakes?
22. Are persons with permanent/temporary disabilities identified and do emergency procedures provide for their safe evacuation?
23. Do all occupants know how to contact security or a communications center to report a dangerous situation (i.e., sticker on phone, "emergency contact button" identified on phone, etc.)?
24. Are emergency phone numbers (i.e., internal/external assistance) posted on all phones?
25. Are the telephone, office detection, alarm, and public address (PA) system supplied with emergency power?
26. In the event of an electrical power failure, is emergency lighting operational for exit access routes, exit enclosures, all rooms with capacity more than 50 people, and normally occupied windowless or underground rooms larger than private office?
27. Is there an emergency generator to provide power to emergency lighting, fire alarm system, fire pump, central control station equipment, and lighting, and is it tested annually?
28. Is there provision for directing occupants to a safe meeting place away from the building when they reach the discharge floor level?
29. Are smoke detectors tied into the building's fire detection/alarm system?
30. Is servicing, maintenance, and testing of smoke-detection systems, fire suppression, and alarms performed by a trained person knowledgeable in the system?
31. Is a passive or mechanical system in place to control smoke in stairwells?
32. Are smoke detectors located in return air ducts of the air conditioning system?
33. Are vertical openings in fire-rated assemblies properly sealed and fire-stopped?
34. Inspect a representative number of openings around piping and other openings in electrical or telephone equipment rooms or closets. Are they plugged with non-combustible material? Are wall and interior finishing materials highly combustible?

35. Is the amount of flammable liquid present at the workplace kept to the minimum amount required for one day's work?
36. Is smoking prohibited, and are other ignition sources excluded from areas where the liquids are present?
37. Are fire doors and shutters in good operating condition and protected against obstructions, including their counterweights?
38. Are fire door and shutter fusible links in place?
39. Are automatic sprinkler system water control valves, and air and water pressure checked weekly/periodically as required?
40. Is the maintenance of automatic sprinkler systems assigned to responsible persons or to a sprinkler contractor?
41. Are doors, passageways, or stairways that could be mistaken for exits appropriately marked "NOT AN EXIT", "TO BASEMENT", "STOREROOM", etc.?
42. Are exit doors side-hinged?
43. Are at least two means of egress provided from elevated platforms, pits, or rooms where the absence of a second exit would increase the risk of injury from hot, poisonous, corrosive, suffocating, flammable, or explosive substances?
44. Is the number of exits from each floor of a building and the number of exits from the building itself appropriate for the building occupancy load?
45. Are exit stairways, which are required to be separated from other parts of a building, enclosed by at least a two-hour fire-resistive construction in building more than four stories in height with a minimum hour-hour fire-resistive constructive elsewhere?
46. If frame-less glass doors, glass exit doors, or storm doors are used as exits, are they fully tempered and do they meet the safety requirements for human impact?
47. Are windows that could be mistaken for exit doors made inaccessible by means of barriers or railings?
48. Are exit doors operable from the direction of exit travel without the use of a key or any special knowledge or effort when the building is occupied?
49. Is a revolving, sliding, or overhead door prohibited from serving as a required exit door?
50. If panic hardware is installed on a required exit door, can the door still open by applying a force of 15 pounds or less in the direction of the exit traffic?
51. Are doors on cold storage rooms provided with an inside release mechanism that will release the latch and open the door even if it's padlocked or otherwise locked on the outside?
52. Where exit doors open directly onto any street, alley, or other area where vehicles may be operated, are adequate barriers and warnings provided to prevent employees from stepping into the path of traffic?

RESPIRATORY PROTECTION AND PERSONNEL PROTECTIVE EQUIPMENT

1. Do any of the work procedures require the use of personal protective equipment (PPE)?
2. Is supply and replacement of PPE recorded?
3. Are records of PPE training and instruction maintained?
4. Does evidence exist that PPE maintenance/storage procedures are followed and are records kept?
5. Is a program for noise evaluations and hearing conservation documented, and have noise hazardous areas been identified, including requirements to use hearing protection in high-noise areas?
6. Are noise-hazardous areas or equipment posted and labeled?
7. Are employees qualitatively or quantitatively fitted for respirators in accordance with industry and regulatory requirements?
8. Are employees trained to understand the limitations of respiratory protection, respiratory protection selection and fitting, and the periodic testing and maintenance requirements for the device?
9. Does the PPE policy address the use of PPE by visitors and outside contractors?
10. Does the facility's PPE program require the use of safety glasses and goggles/face shields to protect against flying chips, chemical splashes, and sparks?
11. Does the facility PPE program require the use of leather/metal mesh gloves (for handling sharp objects), insulated gloves (for handling hot or cold objects), rubber/synthetic gloves (for chemical handling), and dielectric gloves (for electrical hazards)?
12. Does the facility PPE program require the use of rubber/synthetic aprons and clothing (for chemical splashes), insulated clothing (for hot and cold work areas)?
13. Does the facility PPE program require the use of safety shoes/boots (to prevent crushing injuries) and rubber boots (for chemical/liquid handling)?
14. Does the facility PPE program require the use of harness/safety belt/lanyard (cord-rope) for prevention of falls?
15. Are approved safety glasses required to be worn at all times in areas where there is a risk of eye injuries?
16. Are hard hats inspected periodically for damage to the shell and suspension system?
17. Are eye wash facilities and a quick-drench shower within the work area for employees exposed to liquid chemicals and potential splash hazards?

TOOLS–EQUIPMENT, INCLUDING MAINTENANCE

1. Are tools inspected and maintained to ensure design performance and that guards or other safety devices are in place and used?
2. Is the tool free of oil, grease?
3. Are there any visible cracks in the blade or handle, and is the correct size tool for the job used? (Pay particular attention to the striking surface and handle.)
4. Are the surfaces of the jaws worn, damaged, or otherwise out of alignment?
5. Are foot-operated switches guarded or arranged to prevent accidental actuation by personnel or by falling objects?
6. Are manually operated valves and switches controlling the operation of equipment and machines clearly identified and readily accessible?
7. Are all pulleys and belts that are within 7 feet of the floor or working level guarded?
8. Are all moving chains and gears guarded?
9. Are splashguards mounted on machines that use coolant to prevent the coolant from reaching employees?
10. If machinery is cleaned with compressed air, is air pressure controlled and personal protective equipment or other safeguards utilized to protect operations and other workers from eye and body injury?
11. Are fan blades protected with a guard having opening no larger than 1/2-inch when operating within 7 feet of the floor?
12. Are guards in place to prevent employees from coming in contact with rotating or other moving parts or nip points?
13. Are interlocks in place to prevent activation of equipment when guards or electrical panels are lifted?
14. Is all machinery and equipment kept clean and properly maintained?
15. Is sufficient clearance provided around and between machines to allow for safe operations, set up or servicing, material handling, and waste removal?
16. Are equipment and machinery securely placed and anchored, when necessary, to prevent tipping or other movement that could result in personal injury?
17. Do large conveyor systems have a time delay and automatic buzzer and/or lights that operate before the conveyor starts to move?
18. Is there a written machine safeguarding program in place?

Ask to see several major pieces of equipment to verify if points of operation hazards are adequately safeguarded.

Ask to observe machine operators to verify that they are not required to reach into the point of operation to feed or remove parts.

Ask to see several major pieces of equipment to verify if other machine hazards such as moving gears, belts, or chains are adequately guarded.

Ask to see several major pieces of equipment to verify if guards are made of durable materials and free of sharp edges and if the guards totally enclose the hazards and are properly installed and interlocked (when appropriate).

Verify that foot and hand controls are protected against accidental activation.

INSPECTION CHECKLIST FOR PNEUMATIC TOOLS

1. Do the hose connections to the tool and to the compressor fit snugly enough to prevent air leaks?
2. Do all fittings designed to blow compressed air for cleaning have pressure-reduction devices (to less than 30 psig)?
3. Do all hose fittings have safety chains to prevent whipping of the hose if a connection comes loose?
4. Does the air hose have a safety check valve at or near the compressor connection that will shut off or bypass the airflow if a break occurs in the air hose?
5. Are there any visible cracks or defects in the tool housing?

INSPECTION CHECKLIST FOR HYDRAULIC POWER TOOLS

1. Are signs of fluid leakage around hydraulic lines, cylinders, reservoirs, pumps or other system components visible?

INSPECTION CHECKLIST FOR POWDER-ACTIVATED TOOLS

1. Have workers been trained and qualified (licensed or authorized) to operate powder-actuated tools in accordance with manufacturer's instructions?
2. Is powder-activated tool stored in a locked container when not in use?
3. Are powder activated tools inspected for obstruction prior to use?

HOIST AND AUXILIARY EQUIPMENT

1. Is each overhead electric hoist equipped with a limit device to stop the hook travel?

2. Is the rated load of each hoist marked and visible to the operator?
3. Are stops provided at the safe limits of travel for trolley hoist?
4. Are the controls of hoist plainly marked to indicate the direction of travel or motion?
5. Is each cage-controlled hoist equipped with an effective warning device?
6. Are close-fitting guards or other devices installed on hoists to ensure that hoists ropes will remain in the sheave groves?
7. Are nip points or contact points between hoist ropes and sheaves that are located within 7 feet of the floor or work platform guarded?
8. Is the use of kinked or twisted chains or rope slings prohibited?
9. Is the operator instructed to avoid carrying loads over people?

ABRASIVE WHEEL EQUIPMENT-GRINDERS

1. Is the work rest used and kept adjusted to within 1/8-inch of the wheel? Is the adjustable tongue on the topside of the grinder used and adjusted to within 1/4-inch of the wheel?
2. Do side guards on the abrasive wheel cover the spindle, nut, and flange and 75 percent of the wheel diameter?
3. Are bench and pedestal grinders permanently mounted?
4. Are goggles or face shields worn when grinding?
5. Is the maximum RPM rating of each abrasive wheel compatible with the RPM rating of the grinder motor?
6. Are fixed or permanently mounted grinders connected to their electrical supply system with metallic conduit or other permanent wiring?
7. Is each electrically operated grinder grounded?
8. Before new abrasive wheels are mounted, are they visually inspected and ring tested?
9. Are dust collectors and exhaust systems provided on grinders used for dust-producing operations?
10. Are splashguards mounted on grinders that use coolant?

HAZARDOUS MATERIALS AND HAZARDOUS WASTE REQUIREMENT CHECKLIST

1. Do procedures specify requirements for the control of materials subject to deterioration or a "use by" date? Does the organization have documented procedures that conform to legislative requirements, standards, and codes for the storage, handling, and transport of hazardous substances?

2. Are storage areas appropriately segregated, and is secondary containment such as dykes and birms used to capture a percentage of the total volume of chemicals in storage in the area?
3. Is the wall-floor interface sealed to contain fluids in the event of a release?
4. Are the storage areas appropriately protected with barriers to prevent damage from forklifts or other vehicles?
5. Are all containers of HM/HW appropriately labeled, including the date for storage of HW? (Labels must include the necessary precautions regarding the handling storage of the material and its chemical and common names.)
6. Does the facility label all pipes and temporary containers for use or transport of HM with the appropriate labels?
7. Are floor gutters, floor drains, or storm drains protected to prevent the collection of HM/HW in them?
8. Are the floors of HM/HW storage made of nonporous material such as steel or concrete to contain potential spills of HM/HW?
9. Are there cracks or other openings in the floor of the storage container or HM/HW storage area?
10. Is the HM/HW storage area located near streams, rivers, lakes, swamps, or other water bodies that could be contaminated by a release of the material?
11. Are there eyewashes and showers in the HM/HW storage or use area?
12. Is protective equipment worn, including respiratory protection, gloves, eye protection, aprons, boots, as needed depending on the potential exposure risk?
13. Are reactive materials appropriately separated from each other or from water to prevent such an incident?
14. Are chemicals that are known carcinogens, teratogens, or sensitizers used in the facility? Are employees trained to work with these materials, and are appropriate controls in place to eliminate or control the potential for exposure?
15. Is a documented employee hazard communication and community emergency response plan in place?
16. Are controls adequate for the level of risk, and is information such as material safety sheets, inventory/storage records, and storage maps been prepared and have stakeholders been trained, including periodic drills, as appropriate?
17. Are hazardous material requirements reviewed by qualified personnel or consultants to determine if less hazardous material can be substituted?
18. Are tanks and drums appropriately supported, contained, and electrically grounded?
19. Is HM cabinet storage is limited to 60 gallons of Class I or II materials or to 120 gallons of Class III materials?

20. Are compressed gases stored upright with their caps on and secured to prevent falling over?
21. Are ventilation hoods flanged and designed in such a way to maximize capturing efficiency and to draw hazardous material vapors and mists away from the employee-breathing zone?
22. Is there routine maintenance of the ventilation system, and does the operator have a gauge or other visible control monitoring device to document that the system is working in accordance with design requirements?
23. Are oxidizing chemicals stored separately from organic materials?
24. Are flammable materials stored in metal containers?
25. Are acids and corrosives stored separately?
26. If laboratories store more than 10 gallons of Class I or Class II chemicals outside the storage cabinet are they stored in safety cans?
27. Is there mechanism to keep MSDS's (material safety data sheets) up-to-date? Are MSDS's kept in one or more key locations, readily available to supervisors, employees, and first-aid providers (medical personnel) on all shifts?
28. Does the employee sign-off to confirm periodic training on the use of hazardous chemicals in the workplace, MSDS's, physical and health hazards of the chemicals in the work area, and safe work practices, such as storage, disposal use of PPE?
29. Is there an inventory of hazardous waste materials, including quantities and disposal methods names of transporters, manifest, and dates placed in storage?
30. Is all hazardous waste stored and labeled in accordance with regulatory requirements, storage date, and chemical identification?
31. Are licensed documented transporters used for hazardous waste, and are shipping and disposal manifests records readably available?
32. Is employee exposure to chemicals in the workplace controlled to allowable limits?
33. Are spray painting operations done in spray rooms or booths equipped with an appropriate exhaust system?
34. When the contents of pipelines are labeled and identified by name or abbreviation, is the information readily visible on the pipe near each valve or outlet?
35. Does an automatic interlock turn on the local exhaust system whenever the process/equipment is being operated?
36. Is the volume and velocity of air in each exhaust system sufficient to capture the dusts, fumes, mists, vapors or gases and carry them to a collection point or filter?
37. Are exhaust inlets, ducts, and plenums designed, constructed, and supported to prevent collapse or failure of any part of the system?

38. Are clean-out ports of doors provided at intervals not to exceed 12 feet in all horizontal runs of exhaust ducts?
39. Where two or more different type of operations are being controlled through the same exhaust system, will the combination of substances being controlled constitute a fire, explosion, or chemical reaction hazard in the duct?
40. Is adequate make-up air provided to areas where exhaust systems are operating?
41. Is the source point for make-up air located so that only clean air is brought into the area?
42. Where two or more ventilation systems are serving a work area, is their operation such that one will not offset the functions of the other?
43. Do solvents used for cleaning have a flash point of 100°F or more?
44. Are "NO SMOKING" signs posted in spray areas, paint rooms, paint booths, and paint storage areas?
45. Is the spray area kept clean of combustible residue?
46. Are spray booths constructed of metal, concrete, masonry, or other substantial noncombustible material and are they easily cleaned?
47. Are lighting fixtures for spray booths located outside of the booth and is the interior lighted through sealed clear panels?
48. Are the electric motors for exhaust fans placed outside booths or ducts?
49. Are belts and pulleys inside the booth fully enclosed?
50. Is infrared or other heated drying separate from paint spraying operations or is the spray area ventilated before the heated drying process is started?
51. Is a procedure in place that checks all incoming materials to ensure that it meets customer/operations procurement specifications?
52. Is a procedure in place that ensures hazardous materials are appropriately identified and labeled at all stages of the production process?
53. When materials are decanted into smaller containers, are the smaller containers labeled?

MANUAL MATERIAL HANDLING

1. Can the task be altered? (reorganize workplace use of conveyors, deliveries made to point of storage, passageways widened to allow mechanical aids, shelf heights lowered to reduce stretching, etc.)
2. Can the environment be made safer? (level flooring, improved lighting, comfortable temperature, good housekeeping, etc.) Can jobs be rotated to avoid repetition and to prevent over exertion?
3. Are lifting devices available to assist with moving heavy or awkward materials or material storage containers?

4. Has work been organized to eliminate or minimize manual handling operations? Can mechanical systems be introduced?
5. Can the load be improved?(i.e., made smaller, lighter, handles provided, jagged edges smoothed, etc.)

MISCELLANEOUS SAFETY CONCERNS

1. Are emergency lighting and power equipment installed, tested, and maintained where appropriate?
2. Are an adequate number of emergency showers and eye washes installed, accessible, tested, and maintained?
3. Are individuals at risk of falling from work areas, structures, or equipment? What controls are in place to prevent this?
4. Are walking surfaces slippery or wet, increasing the potential for slips and falls?
5. Are make shift ladders used? Are ladders purchased and used in accordance with standards?
6. Is scaffolding used instead of ladders, and does the scaffolding meet the requirements in standards for height, width ratio, footing, toe boards, and guard rails and is the floor of the scaffold covered with a non-skid surfacing material?
7. Are all uprights provided with base plates or prevented in some other way from slipping or sinking?
8. Have any uprights, ledges, braces or struts been removed?
9. Is the scaffold secured to the building in enough places to prevent collapse? Have any ties been removed since the scaffold was erected?
10. Do workers at heights have fall protection and appropriate harness and lanyards?
11. Are warning notices posted to prohibit the use of any scaffold that is not fully erected or boarded?
12. Do tanks and boilers have pressure-relief valves, and are the valves the appropriate size for the pressurized vessels?
13. What is the testing and maintenance schedule for the container-pressure vessel and the relief vessel, and is this schedule adequate for the risk or does it meet minimum regulatory requirements?
14. Are vehicles provided with seat belts?
15. Are all vehicles maintained on a regular schedule?
16. Does the facility have a documented program for the identification and control/elimination of biological hazards (blood-borne pathogens, bacteria, fungus, etc.)?

17. Are initial and periodic evaluations conducted to identify sources of biological hazards, such as cooling towers, biological hoods, ventilation ducts, blood and other biological materials, and processes and exposures?
18. Does the facility have current operating instructions, maintenance, and calibration records for HVAC and control ventilation systems components, including periodic monitoring/testing of the system?
19. Are drivers qualified only after examination of driving records, medical and drug screening, and training?
20. Are required licensed or authorized operators available for equipment such as overhead cranes, forklifts?
21. Does the company vehicle driver selection include a past employer check, license/traffic violations, personal references check, and a follow-up check ride after hiring?
22. Are warning, stop, and speed signs; speed bumps; and blind-corner mirrors included in the in-plant traffic control plan?
23. Do in-plant vehicles have appropriate operational safety equipment such as mirrors, back-up alarms, horns, and flashing lights?
24. Do the truck and tractor–trailer drivers receive special training for handling and transporting hazardous materials?
25. Are blocks in place to prevent movement of vehicles that jacked up for repair?
26. Are inspections and maintenance on cranes, wire ropes and chain slings performed and documented by authorized/certified individuals?
27. Are walk-over stairs and bridges over conveyors available where necessary?
28. Are periodic documented inspections held for freight and passenger elevators and escalators? Are licensed certificates posted near/on the equipment?

Appendix III

Who is Lincoln Electric and What Is the Lincoln Incentive Management System?

The company started in 1895 and is headquartered in Cleveland, Ohio. It manufactures welders and welding supplies. The Lincoln Electric system, according to its own managers, is really not a system but a philosophy. A philosophy started by James Lincoln and his brother, John, who founded the company.

The philosophy, as described in James Lincoln's book and by Lincoln management, centers on six principles:

1. **People:** LE believes that employees are the company's most valuable assets. They must feel secure (like Deming's belief in removing fear), important, challenged, in control of their destiny, confident in their leadership, responsive to common goals, and that they are being treated with integrity; and they should have easy access/lines of communications to management.
2. **Christian ethics:** James Lincoln was a strong believer in the Sermon on the Mount, which established the foundation of the company's culture.
3. **Principles:** Managers must continuously seek the principle involved in any question or evaluation to avoid erratic or improper decisions or actions.
4. **Simplicity:** That is, simplicity in thought, policy, principle, building, offices, assembly lines, personnel practices, and organizational structure.

5. **Competition:** They believe competition is healthy, desirable, and as necessary for successful individuals as it is for business
6. **Customer**: The company exists solely to serve the customer. They see their mission as producing the best products possible to sell at the lowest possible price to an increasing number of customers. Lincoln believes that if this goal is accomplished, the needs of the employees and shareholders will be more than adequately satisfied.

Lincoln has had a guaranteed employment policy that ensures at least 30 hours of employment for those who have worked for the company for more than two years, unless terminated for cause. Part of this policy also requires that employees be willing to transfer if necessary to ensure full employment, and promotions are primarily from within. Vacations are 4 weeks for those with less than 25 years and 5 weeks for those with more than 25 years of service.

Lincoln adjusts pay annually to be competitive with their markets and changes in the Consumer Price Index, and they try to pay on the basis of piecework for the most part. However, piecework rates can be 40–50% more than what a skilled operator or mechanic earns. Bonuses have averaged 90–110% of annual pay since 1950.

The bonus has been as high as 120% of pay and as low as 55%. At the same time the company maintains all other traditional fringe benefits such as life insurance, health insurance, paid vacations, and an annuity plan. Employees with more than one year of service also have a stock purchase plan. The stock must be sold back to the company if the employee leaves, and employees hold about 80% of the stock.

Given this pay structure and evaluation system, first-time quality is expected from everyone. If a product does not conform to specifications, then the team that built it corrects the mistake on its own time. Warranty claims are very low: less than 1 in 1,000 units produced.

The company operates on the principle of management by exception, using the fewest possible layers of management and the fewest number of managers as possible to achieve desired results.

Income Distribution averages

- Materials 46%
- Taxes 9%
- Shipping 5%
- Operating expenses 6%
- Dividends 3%
- Reinvestment in the business 5%
- Annuity 2%
- Wages 13%
- Bonus 11%

Example of a modified Lincoln Electric Evaluation & Bonus Plan Payoff*

Output. This element rates how much productive work you produce. It also reflects your willingness to aggressively work with management and clients to generate new work and deliver high-quality projects. As a more quantitative measure, it considers your utilization rate against the plan and recognizes your attendance record.

Quality. This element rates the quality of the work you do. It evaluates your diligence in providing clients with quality services and documenting adherence to the your company's quality management system. It also reflects your success in eliminating errors and in reducing rework and ability to execute procedures by using the your company's quality management check sheet and quality system

TABLE 1. Sample Bonus Pay Out

Needs:	Total Salary cost for the year				
	Pay-out factor that is equal to pay-out pool (in $) divided by total salary cost Individual performance factors The Inverse of the average performance factor for the pool participants Pool = 30% of gross Income Performance evaluations every 6 months are averaged for the end-of-year bonus pay-out				
Example:	Revenues = **$1.5 M** Profits = **$225 K** Bonus pool = 30% of profits or 225 = **$67.5 K** Pay-out factor = $67.5/$521 K = **0.13**				
Staff	**Salary**	**Avg. Performance Score**	**Pay-Out Factor**	**Inverse Avg. Score Performance**	**Bonus**
Mike	$75,000	0.95	0.13	1.15	$10,652
Carroll	$70,000	0.80	0.13	1.15	$8,372
Rachel	$65,000	0.75	0.13	1.15	$7,288
Debbie	$58,000	0.85	0.13	1.15	$7,370
Steve	$58,000	0.95	0.13	1.15	$8,237
Bill	$50,000	0.90	0.13	1.15	$6,728
John	$50,000	0.85	0.13	1.15	$6,354
Karl	$45,000	0.90	0.13	1.15	$6,055
Bob	$25,000	0.95	0.13	1.15	$3,551
Jane	$25,000	0.80	0.13	1.15	$2,990
	Avg. Performance Score	0.87		TOTAL:	$67,596
	Inverse of Average Performance Score		1.149		

* Used by author while president of certified environmental consulting.

as it relates to project work. In addition, this element reflects your own personal commitment and adherence to your company's safety policies

Dependability. This element rates how well management can depend upon you to accomplish work without supervision. It also rates your ability to supervise yourself, including your work safety, care of your company's customers, quality, your orderliness, care of equipment, and the effective use of your skills.

Ideas and Cooperation. This element rates your cooperation, ideas, and initiative. Your company depends on all employees to market and sell its services and to help your company design and deliver new products and services. New ideas and new methods are important to your company in your continuing effort to reduce costs, increase output, improve quality, work safely, and improve the company's relationship with its customers. This element credits you for your ideas and initiative used to help. It also rates your cooperation-how you work with others as a team. Factors such as your attitude toward supervision and your co-workers and the company; your efforts to share your expert knowledge with others; and your cooperation in installing new methods smoothly are rated.

Appendix IV

Comparison of Safety, Environmental and Quality Management Systems

ISO 14001 Standard	ISO 9001 Standard	Draft OHSAS 18001
4.0 General Elements	4.1, 4.2	4.0 OH&S Management System Elements
4.1 General Environmental Requirements	4.1 General Quality Requirements	4.1 General OH&S Requirements
4.2 Environmental Policy	4.2 Quality System	4.2 OH&S Policy
4.2.1 Environmental Planning Aspects		4.2.1 General OH&S Planning Aspects
4.3 Planning	4.3 Quality System	4.3 Planning
4.3.1 Environmental Aspects	4.2 Quality System	4.3.1 Planning for Hazard Identification, Risk Assessment and Control
4.3.2 Legal and Other Requirements		4.3.2 Legal and Other Requirements
4.3.3 Objectives and Targets	4.2 Quality System	4.3.3 Objectives
4.3.4 EMS System Programs		4.3.4 Occupational Health and Safety Management System Programs

ISO 14001 Standard	ISO 9001 Standard	Draft OHSAS 18001
4.4 Implementation and Operation	4.2 Quality System	4.4 Implementation and Operation
	4.9 Process Control	
4.4.1 Structure and Responsibility	Mgt. Responsibility	4.4.1 Structure and Responsibility
	4.1.2 Organization	
4.4.2 Training Awareness and Competence		4.4.2 Training Awareness and Competence
4.4.3 Communication		4.4.3 Consultation and Communication
4.4.4 Environmental Management System Documentation	4.2.1 General	4.4.4 Documentation
4.4.5 Documentation and EMS Record Control	4.5 Document and Data Control	4.4.5 Documentation and Data Control Records

Record Control		Records
4.4.6 Operational Control	Quality System Procedures	4.4.6 Operational Control
	Contract Review	
	Design Control	
	Purchasing	
	Customer Supplied Products	
	Product Identification	
	4.9 Process Control	
	4.15 Handling, Storage, Packaging, Preservation and Delivery	
	4.19 Servicing	
	4.20 Statistical Techniques	
4.4.7 Emergency Preparedness and Response		4.4.7 Emergency Preparedness and Response
4.5 Management Review: Checking and Corrective Actions	4.2	4.5 Management Review: Checking and Corrective Actions
4.5.1 Monitoring and Measurement		4.5.1 Performance Measurement and Monitoring

Record Control		Records
4.5.2 Nonconformance and corrective and preventive actions	4.13 Control of Nonconforming Products	4.5.2 Accidents, Incidents, Nonconformance and Corrective and Preventive Actions
	4.14 Corrective and Preventive Action	
4.5.3 Records and Record Management	4.16 Quality Records	4.5.3 Records
4.5.4 Environmental Management System Audit	4.17 Internal Quality Audits	4.5.4 Audit
4.6 Management Review	4.1.3 Management Review	4.6 Management Review

Appendix V

SAIC; The largest Employee Owned Company in The World, #347 in Fortune 500 Listing

SAIC was founded 30 years ago by J. Robert Beyster. Prior to starting SAIC he had managed a 200-person research group at a large corporation. He became frustrated when the profits generated by his team were directed towards other priorities reflecting a strong management focus on outside investors. Mr. Beyster started SAIC with the goal of "creating an environment for technical people to solve important problems for their customers, an environment where the priorities of the company would be driven by those doing the work."

When he incorporated SAIC he only took 10% of the stock for himself, and gave the rest of the ownership to those employees who would be building it. SAIC generated $250,000 in revenues the first year, and through its employee ownership generated over $5 billion in revenues last year, and are now ranked 347 in the Fortune 500 listing.

J. Robert Beyster is still Chairman and CEO of SAIC, and has developed this amazing technology company on a belief that when people work hard and take risks to make something work, they want to "own it". SAIC uses virtually every mechanism available to put ownership in the hands of the employees, and has generated more that one millionaire as a result of this philosophy.

SAIC expertise expands business and technology horizons in fields as diverse as energy, environmental health and safety management, health and national security. SAIC helps companies integrate their quality, and environmental health and safety programs, and to better recognize and control process risks.

Their cutting-edge simulations help energy companies produce more oil from underground reservoirs and NATO forces prepare for peacekeeping missions. Their scientists investigated new pharmaceuticals, while their engineers improved the reliability and availability of radar systems used to control aircraft landings. Their Year 2000 services prepared all kinds of enterprises to meet the future with confidence. During conflict situations, SAIC staff rapidly deploys into the field with military personnel to set up theater command and control systems.

SAIC has retained its small company flexibility and entrepreneurial edge. Its dynamic, decentralized organization allows it to quickly shift resources where they will best meet the clients changing needs and create a competitive advantage. There is strong synergy between the company's different technical fields who span an unusually diverse range of disciplines to create the best, end to end solutions for their clients.

Key business solutions for Industries and Governments Include:

- Information Protection
- Outsourcing and Desktop Computing
- E-Commerce and Supply Chain Management
- Program Management and Systems Integration, including Environmental Health and Safety
- Software Development

Some of SAIC's Key Vertical Markets Include:

- Telecommunications
- Health Care
- Environmental Health and Safety
- National Security
- Energy
- Transportation & Logistics
- Space
- Law Enforcement

Employee ownership has in no small way contributed to SAIC's success, and spectacular growth over the last 30 years. The employee ownership system helps them to recruit and retain noted experts and helps to motivate them to perform at their best for their customers. Employee ownership has made SAIC a company that stands out among its competitors. In fact, many of the competitors they have had in the last 30 years have disappeared.

References and Additional Reading

GENERAL ORGANIZATIONAL MANAGEMENT

Clark, K., 1999. "Why it pays to Quit: Career Guide 2000," *U. S. News & World Report* (November 1999) p. 78.

Drucker, P. F., 1989. *The New Realities*, Harper Row.

Drucker, P. F., 1992. *Managing for the Future: The 1990s and Beyond*, Truman Talley/Plume Books.

Peters, T. and Waterman, R., 1982. *In Search of Excellence*, Harper Row.

Quinlan-Hall, D. and Renner, P., 1940. *In Search of Solutions, Sixty Ways to Guide Your Problem Solving Group*, PFR Training Associates Limited.

QUALITY MANAGEMENT

Automobile Manufactures QS 9000, *Quality Management Standard.*

_____. 1999. "Boosting the Bottom Line Over Time: Study Shows TQM Pays Off in Breyfogle, F. W., 1999. *Implementing Six Sigma*, John Wiley & Sons.

Crosby, P. B., 1979. *Quality is Free*, McGraw Hill.

Crosby, P. B., 1994. *Completeness, Quality for the 21st Century*, Plume.

Deming, W. E., 1982. *Quality, Productivity, and Competitive Position*, MIT Center for Advance Engineering Study.

Hiam, A. and The Conference Board, 1992. *Closing the Quality Gap: Lessons from Americas Leading Companies*, Prentice Hall.

Higher Stock Prices, and Profitability, *Competitive*, Vol. 8, Number 2, Summer 1999; American Society for Quality, p. 1–4.

Hoerl, R. W., 1996. "Six Sigma and the Future of the Quality Profession," *Quality Progress* (June 1996) pp. 35–42.

ISO 9000, 1994.

Kanholm, J., 1994. *ISO 9000 Quality System, 1994 Standards, 2nd Edition*, AQA Co.

Mil Standard 882, Rev D System Safety Program Requirements.

ENVIRONMENTAL MANAGEMENT

Cascio, J., Woodside, G. and Mitchel, P., 1996. *ISO 14000 Guide*, McGraw Hill.

Chemical Manufactures Responsible Care Program.

Epstein, M. J., 1996. *Measuring Corporate Environmental Performance*, Irwin Professional Publishing.

ISO 14000, 1996.

OCCUPATIONAL HEALTH & SAFETY MANAGEMENT

Bird, F. and Germain, G., 1985. *Practical Loss Control Leadership*, Institute Publishing.

Daniels, A., 1989. *Performance, Improving Quality, Productivity Through Positive Reinforcement*, Performance Management Publications.

Draft Australian/New Zealand Occupational Health and Safety Management System, 1996.

Dyjack, D. T. and Levine, S. P., 1995. Development of an ISO 9000 Compatible Occupational Health Standard: Defining the Issues. *Amer. Ind. Hyg. Assoc. J.* 56: 599–609.

Dyjack, D. T., Levine, S. P., Holtshouser, J. L. and Schork, M. A., Comparison of AIHA ISO 9001-based Occupational Health and Safety Management System Guidance Document With a Manufacturer's Occupational Safety and Health Assessment Instrument. *Amer. Ind. Hyg. Assoc. J.* 59: 419–429.

Geller, S., 1998. *Working Safe*, CRC Press.

Krause, T. R., 1995. *Employee Driven Systems for Safe Behavior*, Von Nostrand Reinhold.

Levine, S. P. and Dyjack, D. T., 1995. Development of an ISO 9000 Compatible Occupational Health Standard—II: Defining the Potential Benefits and Open Issues, *Amer. Indus. Hyg. Assoc. J.* 57: 387–391.

Levine, S. P. and Dyjack, D. T., 1997. "Critical Features of an Auditable Management System for an ISO 9000-Compatible Occupational Health and Safety Standard." *Amer. Ind. Hyg. Assoc. J.* 58: 291–298.

Occupational Health and Safety Management System, BS 8800, 1986.

OHSAS 18001 Draft Occupational Health and Safety Management Systems-Specifications, British Standards Institute, National Standards Authority of Ireland.

Redinger, C. F. and Levine, S. P., 1996. "New Frontiers in Occupational Health and Safety: A Management Systems Approach and the ISO Model" *AIHA Policy Monograph.*

Redinger, C. F. and Levine, S. P., 1997. OHS Performance Measurement Systems: Analysis of Industry Practices, and their Relationship to OSHA's Performance Measurement System and the Government Performance and Results Act of 1993. OSHA Contract J-9-F-3-0043.

Redinger, C. F. and Levine, S. P., 1997. Implementation of an ISO 14001-Based OHS Management System — A Pre-Registration Gap Analysis. NASA Lewis Research Center Contract C-77958-F.

Redinger, C. F. and Levine, S. P., 1999. "The Universal Assessment Instrument for Occupational Health And Safety Management System" *AIHA Publications*.

Index

ABB, 181
ABC analysis:
  exposures, implications of, 167–169
  problem-solving, 169
  steps of, 163–167, 169–174
Absenteeism, 40
Acceptance, 5, 45, 143
Accident Injury Program, 27
Accidents:
  cause of, 32, 39, 137. *See also* Multiple root cause analysis
  cost of, xxxiv, 27, 34, 36
  investigation of, 189
  losses, 21
  lost-time, 142
  nonconformance and, 41
  prevention of, *see* Safety management
  serious, 27
Accountability, 5, 7–8, 36, 41, 60, 68, 231–232
Action plans:
  in implementation phase, 73–74
  incident investigations, 198
  integrated management systems (IMS), 23–26
  management commitment, 55–56
  management reviews, 214
  measurement and evaluation, 185
  operational control, 130–131
  risk assessment process, 111
  in safety management, 43
Administrative controls, 123, 128
Alcoa, 18
Allied Signal, 22, 181
American Industrial Hygiene Association, 37
American Society for Quality's Audit Committee, xxxii
Amputations, 37
Annual report, 51
Antecedents, job-task analysis, 162, 168. *See also* ABC analysis
Appraisals, 59
Archives, contents of, 71
Asbestos removal, 106
Asthma, 37
At-risk behavior, 167
AT&T, 99
Attitude, job-task analysis, 162
Auditor:
  qualifications of, 151
  role of, 212
Audits:
  attributes, overview, 147, 210–211, 229–259
  benchmark, 213
  cross-facility, 208, 210, 217
  due-diligence, 210
  financial, 210
  follow-up, 184, 210, 217
  frequency of, 151, 212
  guidelines for, 150–151
  hazard risk identification, 116–117

Audits: (*Continued*)
internal, 151, 211–214
management system, 210–211, 217
operational control, 114, 116
planned, 138
process discussion, 148–150
program, 147–148
purpose of, generally, 4–5, 8, 60, 214
regulatory compliance, 210
reports, 151, 208
risk assessment, 111
routine, 145
safety management, 33–34, 42, 49–50
schedule, 151, 211
third-party, 114, 210, 217
Automobile manufacturers, 13
Avery Dennison, 181

Barrier analysis, 154
Behavior, in job-task analysis, 162. *See also* ABC analysis
Behavioral-based programs, 221–223
Behavioral management, 168–169
Behavioral safety management, 140–144, 194
Behavioral Safety Technology (BST), 140–141
Behavior sampling, 146
Benchmarking:
implications of, 4, 7, 13, 146
management review, 214
performance measurement and, 175–176
Best practices, 6–7, 146
Beyster, J. Robert, 287
Bissett, Sandra, 121
Black Belts, 180–181
Blame:
avoidance of, 192
culture of, 188
implications of, 159, 174
Bonuses, 225
Bottom-line performance, 216
Breakdowns, 64, 183
BS 5750, 216
BS 7750, 216
BS 8800, 2, 216
Budget:
indirect costs and, 35–36
process failure and, 5
work center, 98
Business and organizational QEH&S risk, 81–83
Business Charter for Sustainable Development (BCSD), International Chamber of Commerce, 134
Business development, 226
Business opportunity risk, 227–228
Business risk, categories of, 86

Calibration, 212
Cancer, occupational, 37
Capital, 84, 134
Cardiovascular disease, 37
Carelessness myth, 190–191
Cash flow, 95
Certification, 61, 124
Change Analysis, 60, 110, 154
Change orders, 118
Charge-back-to-work center, 5
Charting, nonconformances, 42
Check sheets, benefits of, 14, 74–79, 146, 152, 163, 194. *See also* Checklists
Checklists:
hazardous materials/hazardous waste requirements, 274–277
supervisory, 32
tool safety, 273–274
Chemical hazards, 35–37, 52. *See also* Hazardous materials/Hazardous waste (HM/HW)
Chemical manufacturer's responsible care (CMRC) program, 13
Chemical Manufacturers Association (CMA), 216
Ciba-Geigy, 22
Civil actions, 40
Climate culture review, 17, 165
Clusters, defined, 175
Coaching, 14, 126
Commitment:
continuous improvement and, 20–21, 208
management, *see* Management commitment and policy, 2, 19–20
training and, 2, 49, 52
Common Cause, 182
Communications:
audit attributes, 234–235
breakdowns, impact of, 142
employee hazards, 125, 129
in implementation phase, 64–66, 217
incident investigations, 195–197
management commitment and, 49–50
in risk assessment, 98–99, 107–108
in risk management, 93
in safety management, 33–35
Community support, 226
Compensation liability, 40
Compensation packages, 41, 225
Competition, business risk and, 86
Complaints:
implications of, 6

measurement and evaluation of, 137
records of, 115
Compounded values, 84
Conflict resolution, 63
Conformance, significance of, 22
Consequences, job-task analysis, 162–163, 169. *See also* ABC analysis
Consultants, role of, 63
Consultation, employee, 64
Consulting Multiplier/Utilization Cost/Profitability model, 219
Contingency plans, 4, 69, 87, 111, 218
Continual process improvement, 4
Continuous improvement:
commitment and, 20–21, 208
implications of, generally, 15, 17, 165, 184
management review, impact on, 207–214
measurement and evaluation, 174
Plan-Do-Correct model (Deming), 226–227
Contractors:
certification process, 61
management and control of, 68–69, 120, 218, 254–255
training, 69
Contract review, 118–120
Control, *see specific types of control*
Corporate quality, sustainability challenge, 20–21
Corrective action:
audit attributes, 243–244
documentation, 184
follow-up, 217
implications of, generally, 4–6, 8, 14, 21
management review of, 209–210
measurement and evaluation process, 133, 139, 145
nonconformance/incident investigations, 195–196
risk management, 42, 102–103
Cost accounting, 99
Cost-benefit analysis, 5, 83–85, 105, 134, 217
Cost control:
in safety management, 39–42
strategies for, 220
Cost, *see specific types of costs*
assessment of, 54
categories of, 135
control, *see* Cost control
Criminal liability, in safety management, 34–35, 40
Critical incident recovery plan, 128–129
Critical tasks:
inspections and, 152
job-task observation and analysis, 153
Critiques, 14
Crosby, Phillip, 2, 5, 22, 29, 45, 168, 216–217
Cross-facility audits, 210, 217
Customer, generally:
complaints, 6, 189, 213
identification of, 136
retention, 227
satisfaction, 136–137
Cycles, defined, 176

Daniels, Aubrey, 140, 168, 171
Data collection:
behavior safety management, 142–143
performance measurement, 175, 183
Death, traumatic, 35, 37
Debugging, functional, 45
Decision symbol, in process flow diagrams, 183
Deere & Co., 18–19
Deming, W. Edwards, Dr., xxvii, 2, 9, 29, 45, 61, 136, 176–177, 216–217
Deming's 14 principles, xxv–xxvi, 8
Deming-Shewart continuous improvement, 17
Department of Energy, xxvii, 139
Department of Transportation (DOT), 40, 125
Depreciation, 84, 88
Dermatological problems, 37
Design:
controls, 8, 45, 121, 218, 244–248
management, role in, 47–48
Diagnostics, 45
Direct costs, 4–5, 84, 133
Discounted cash flow, 88, 98
Discounted value, 84–85
Distribution, business risks and, 86
Documentation:
audit attributes, 212, 240–243
implementation phase, 60, 69–72, 218
importance of, 4, 8
nonconformance, required elements of, 189–190
in risk assessment, 98, 105
in safety management, 33, 36, 38, 42
performance measurement, 184
Drucker, Peter, 23
Due diligence, 51, 210
DuPont, 18–19, 31

EEO, 225–226
EH&S manager, role of, 213
Eighth Point (Deming), 136
Electrical safety, 261–263
Elimination/substitution, 123
Emergency preparedness, 8, 69, 128–129, 218, 255

Emergency response plans, 4, 61, 87, 128–129
Employee(s):
blaming, 159, 174
bonuses, 225
compensation packages, 225
evaluations, 225
involvement, *see* Employee involvement
retention, 40
wellness, 60
Employee assistance programs (EAPs), 38
Employee involvement:
implementation phase, 64, 217
importance of, 14, 33
performance measurement and, 167
Employer:
liability, 39
role in safety management, 38–39
Enforcement, 10, 97
Engineering:
controls, 117–118, 121, 123, 244–248
investigation questions, 201–202
review check sheet, 76
Enterprise Risk, 2
Environmental health and safety program, audit attributes:
electrical safety, 261–263
life-fire safety, 268–270
lock-out/tag-out procedures, 266–268
manual material handling, 277–278
medical and occupational health, 260–261
miscellaneous, 278–279
permit system, 263–266
respiratory protection and personal protective equipment, 271
tools and equipment, 272–274
Environmental Protection Agency (EPA), xxxii, 34, 36, 39, 216
Environmental risk liability, 40
Environmental sampling, 146
Equipment/machinery:
investigation of, generally, 192
investigation questions, 205–206
review check sheet, 79
European Union, safety requirements, 29
Evaluation, *see* Measurement and evaluation
control procedures, 151–152
reports, 145
Exposure evaluation, in risk assessment, 104–105, 109
External communication, 66
External customers, 11, 14
Eye loss, 37

Facilitator, of risk management program, 89–90, 102
Fact-finding, in incident investigation, 191–192
Failure, causative factors, 3–4. *See also* Process failure
Failure Mode Effects Analysis (FMEA), 60, 121, 140, 181, 183
Fault tree analysis, applications of, 96, 116, 139, 163
Feedback, importance of, 143, 161–163, 167, 175
Financial Accounting and Standards Board (FASB), 97
Financial resources, implementation phase, 63
Financing:
post-loss, 89
pre-loss, 89
risk, 88–95
Fire safety, 268–270
Fish bone diagrams, 61, 85, 154, 173
Fixed costs, 143, 220
Flow charting, 60, 66, 183
Follow-up:
to incident investigation, 193
management reviews, 210
risk assessment, 108, 111
Ford, 13, 98–99
Fractures, 37
Fragmentation:
implications of, generally, 31
process control, 11, 19
risk management, 87
Fraud, 15, 166
Future values, 84, 88

Gap Analysis, 57, 62–63, 73–74, 217
Geller, Scott, 140
General Electric (GE), 21–22, 179–181
Gilbert, Thomas, 161, 168
Globalization, impact of, xviii, xxiii
GM, 13, 97
Goals:
improvement, 9
Integrated Management Systems (IMSs), 10
in job-task analysis, 162
Goal-setting, importance of, 162
Graphs, in job-task analysis, 162
Greene, Bill, 27

Hazard risk identification:
audits, 116–117
hazard analysis and systems safety tools, 116
illness/injury records, 115
inspection, 115
job-task analysis, 115
risk assessment, 115
tools/resources used, 114–115

Hazard-Operability (HAZ-OP), 61, 96, 103, 116
Hazardous materials/hazardous waste (HM/HW):
  audit attributes, 250–254
  disposal procedure recommendations, 125
  emergency response and contingency planning, 69
  implementation phase, 218
  indirect costs, 35–36, 38
  operational controls, 124–125
  risk assessment, generally, 47–48, 55, 101, 124–125. *See also* Hazardous risk assessment
  safety guidelines, 38, 124
  spill response plans, 129–130
  storage recommendations, 124–125
  transport recommendations, 125
HAZCOM, 131
Hearing loss, 37
Hold harmless contracts, 89
Human Compliance (Gilbert), 161
Human resources, implementation phase, 63

Illness:
  investigation of, 189
  statistics, 19
Implementation:
  action plan, 73–74
  audit attributes, 233
  communications, 64–66, 217
  components, overview of, 217–218
  contingency planning, 69, 218
  contractors, management and control of, 68–69, 218
  design control, 218
  document control, 69–72, 218
  effective, factors in, 58
  emergency preparedness, 69, 218
  employee involvement and consultation, 64, 217
  financial resources, 63
  human resources, 63
  integration, 63
  management commitment and, 52–53, 216
  and Operational or Risk Control, 2–4
  operations and, 58–63
  physical resources, 63
  procedures, 66–68
  process inputs, process risk analysis, 74–79
  recordkeeping processes, 72–73, 218
  training, 218
  vendors, management and control of, 68–69, 218
Incident investigations:
  action plan, 198
  analysis, 190
  audit attributes, 256–257
  carelessness myth, 190–191
  communications, 195–197
  customer complaints, 189
  equipment/machinery, 192
  fact-finding processes, 191–192
  follow-up, 193
  interview techniques, 192
  key assessment considerations, 191–193
  learning from 195–197
  objective of, 191
  person involved, 192–193
  policy, 187–188
  principles, 193–195
  process policy, 197
  program synergism, 197
  purpose of, 5, 187–188
  questions for, 198–206
  recommendations, 193
  required elements of, 189
  time frame, 193
Incidents:
  investigation of, *see* Incident investigations
  minor, 27, 166
  new employees and, 166
  recordkeeping, 73
Indemnification, 89
Indirect costs, 4–5, 7, 84, 35–37, 133
Individual development plans (IDPs), in training, 127
Inflation, 83–84
Information management, 70
Information system (IS), 134
Initial review:
  operations, 62–63, 73–74, 217
  risk assessment, 100–101
Injuries:
  behavioral safety management, 142
  cost of, 36
  measurement, in investigation of, 189
  occupational, 37
  on-the-job, 4, 19
Injury-Illness Prevention Program, 28
Inspections:
  audit attributes, 237–240
  contractors/suppliers, 120
  documentation, 184
  hazard, guidelines, 125, 147
  in implementation phase, 60, 64, 73
  importance of, 4–5, 115
  management review, 209, 211
  measurement and evaluation, 144–153
  measuring techniques, 146

Inspections: (*Continued*)
planned, 138–139
routine, 145, 152
Insurance:
coverage, 63, 93–95
integrated risk transfer, 88–95
risk management, 175
traditional, 89
Integrated Management Systems (IMSs), 10, 23
Integrated risk management process:
potential participants in, 94–95
risk transfer plan, 93–94
Integration, implementation phase, 63
Internal audits, 151, 211–214
Internal communication, 66
Internal customers, 11, 14
Internal rate of return (IRR), 84, 88, 95
International Labor Organization, 9
Interviews, incident investigations, 191–192
Inventory:
assessment forms, 146
critical parts, 139
development of, useful tools for, 139–140
importance of, 5
Investigation, *see specific types of investigations*
ISO, generally:
emphasis of, 16
establishment of, 8
popularity of, xxxiv
purpose of, 8
ISO 14000, 216
ISO 14000/9000, harmonized quality and environmental health and safety management systems (QEH&S), 1–8
ISO 14001, compared with ISO 9001 and OHSAS 18001, 284–286
ISO 9000:
components of, generally, 216
documentation, four tiers of, 71–72
ISO 9001, compared with ISO 14001 and OHSAS 18001, 284–286
ISO 9001/14001 harmonized management systems, 8–22

Japanese management, 173–174
Job descriptions, 232
Job procedures, written, 66, 126
Job-task analysis:
ABC analysis, 163–174
antecedents, 162
behavior *vs.* attitude, 162
benchmarking, 175–176
benefits of, 158, 217
consequences, 162–163
hazard risk identification, 115
observation checks, 156
performance analysis, 156, 158
performance management, 154, 158–162
performance of, generally, 154
procedures, 155
process control instructions, review of, 154
process management, implications for, 154
purpose of, 4, 13, 21, 99, 146, 153
QEH&S shared system, 174–175
seeing *vs.* observing, 155–156
Six Sigma, 179–183
statistical process control (SPC), 174, 176–179, 181–182
Johnson, William, 96
Juran, xxvii, 2, 9, 29, 45, 176, 181, 216–217

Kaizan, 173–174
Kent, Marthe, 29
Knee-jerk reactions, 177, 181
Krol, John, 18

Labor-management committees, 64
Labor-management relations, 226
Labor organizations, 11
Lange, Scott, 89
Leadership, *see specific types of management*
behavioral-based programs, 221–223
business development and, 226
community support, 226
compensation packages, 225
culture and, 220–221
EEO, 225–226
employee evaluations/bonuses, 225
labor-management relations, 226
lobbying, 226
management control, 223–224
marketing, 226
training/staff development, 224–225
Legislation, implications of, 49, 83, 117
Lessons learned, shared, 15, 33, 49–50, 65, 70
Liability:
criminal, 34–35, 40
employer, 39
environmental risk, 40
integrated risk transfer plans, 94
product, 93, 97
risk assessment, 97
Liberty Mutual, 84, 93, 144
Liberty Mutual Savings Equation, 20
Life cycle, *see* Product life cycle
cost categories, 135
product, 97–98
risk evaluations, 45

Life safety, 268–270
Lincoln Electric:
  generally, 7, 136, 159, 161, 168, 172
  Lincoln Incentive Management System, 280–281
Lineman, Judy M., 93
Line management, 45, 127
Lobbying, 226
Lockheed Martin, 181
Lock-out procedures, 266–268
Loss and accident-incident analysis, 213–214
Loss control, 197
Lung disease, 37

Maintenance:
  audit attributes, 256
  importance of, 8
  investigation questions, 203
  and operating costs, 84, 88
  and repair, 124
  recordkeeping, 212
  and repair, 124
  review check sheet, 77–78
  review and improvement, 4
  standard procedures, 67–68
Malcolm Baldrige Award, 18
Management, *see specific levels of management*
  commitment of, *see* Management commitment
  control, 223–224
  investigation questions, 204–205
  involvement, 14, 33
  job influences, 51–53
  review check sheet, 78–79
Management by objectives (MBO), 135–136
Management by Wandering Around, 171
Management commitment:
  action plan considerations, 55–56
  authority, 50–51, 231–232
  in communication, 49–50
  importance of, 44–48, 208, 216
  job influences, 51–53
  objectives and targets, 53–55
  policy and, 48–49, 229–231
  responsibility, 50–51, 231–232
Management Oversight Risk Tree analysis (MORT), 60–61, 96, 116, 121, 140, 154, 183
Management Program, 8
Management reviews:
  action plan, 214
  audits, 210–213
  continuous improvements, 207–214, 257–259
  corrective action, 209–210
  importance of, 4, 50, 218–219
  operational control and, 114
  performance measurement, 184
  preventive action, 209–210
  purpose of, 20, 137–138
Management system, elements of, 4–6
Marketing, 226
Maslow's Hierarchy of Needs, 171–172
Materials, generally:
  handling, guidelines for, 8, 250–254, 277–278
  investigation questions, 202
  operational controls, 124
  review check sheet, 76–77
Material safety data sheets (MSDS), 125, 129
Measurement and evaluation:
  action plan, 185
  audit attributes, 236–240
  goals of, 4, 137–140
  inspection, 144–153
  job-task observations and analysis, 153–184
  performance/behavioral safety management, 140–144
  purpose of, 132–133
Medical health, 260–261
Medical surveillance programs, 60, 130, 149–150
Microsoft, 20, 89
Mid-level management, 194
Milliken Corporation, 17
Mini-audits, 33
Mitigation process, in risk assessment, 101–102, 111
Monitoring, 8, 38, 58, 143, 146, 152–153
Motorola, 179, 181
Multinational corporations, 13, 35
Multiple root cause analysis:
  benefits of, generally, 5, 14, 49, 97, 108, 137, 217
  performance measurement and, 174
  process failure/incident investigations, 190–191, 195–197
Multiplier, 219

National Aeronautics Space Administration (NASA), xxxiii, 139
Near-misses, investigation of, 189
Negative management, 140
Neurotoxic illness, 37
New employees, 64, 166
New products, introduction of, 52
NFPA, 35
Nonconformance:
  cost of, generally, 7, 22

Nonconformance: (*Continued*)
  customer and, 4–5
  evaluation, 8
  human error and, 142
  incident investigation and, *see* Incident investigation
  management review of, 151
  multiple root causes, 97, 195
  reports, 15, 49
  risks, 6

Objectives:
  audit attributes, 233–234
  implementation phase, 58
  integrated QEH&S risk management model, 21–22
  management commitment, 53–55
Observation:
  behavioral, in performance measurement and evaluation, 144
  significance of, 5
  job, 4, 15, 34, 42, 60, 64
  job-task analysis, *see* Job-task analysis
Occupational disease, 37
Occupational health, 60, 260–261
Occupational Health and Safety management system, 9
Occupational Safety and Health Administration (OSHA), 13–14, 27, 40, 216
OHSAS 18001, compared with ISO 9001/14001, 284–286
Operational business risks, 86
Operational control:
  action plan, 130–131
  audit attributes, 233, 244
  elements of, 112–113
  follow-up, 114
  hazard risk identification, 114–117
  importance of, 8
  management review, 114
  implications of, 124–125, 218
  process flow, 113–114
  process improvement, risk reduction, 122–125
  purchasing, 117–122
  risk assessments, 114
  training, 125–130
Operational risk, 227–228
Operations, in plan implementation:
  initial QEH&S review (Gap Analysis), 62–63, 73–74, 217
  process design and systems safety review, 60–61, 73
  procurement, 61
Operations management:
  measurement and evaluation, 133
  role of, generally, 7, 40
Operations managers, role of, 31
ORC, 84, 93
Organizational culture, implications of, 62, 220–221
OSHA requirements, 26–29
Out-of-the-box thinking, 180
Overhead expenses, 38

Packaging, 8, 250–254
Pareto chart, 169–170, 182
Partial conformance, 16
Payback period, 133
People, business risks and, 86
Performance management:
  behavioral-based, 21, 158–162
  importance of, 168
Performance Management (Daniels), 168, 171
Performance measurement:
  categories of, 161
  and evaluation, 140–144
Performance observations, 13
Permissible exposure limit (PEL), 38
Permit system, 73, 213, 263–266
Personal protective equipment, 123, 271
Peters, Tom, 140
Peterson, Dan, 140
Physical resources, implementation phase, 63
Plan-Do-Correct model (Deming), 226–227
Planning:
  compliance in, 54
  importance of, 2, 45, 51
  management commitment, 52–53, 216
  risk assessment, 102–103
Plant, maintenance and repair guidelines, 124
Pneumoconiosis, 37
Point People, 136
Policy and procedures manual, 47, 54, 59, 70–71
Present value, 84–85, 88
Preventive action:
  audit attributes, 243–244
  documentation, 184
  follow-up, 217
  implications of, generally, 6, 13–14, 22
  management review, 209–210
  measurement and evaluation, 139, 145
  nonconformance/incident investigations, 195–196
  risk assessment and, 102–103
Principle of Critical Few, 193
Principle of Definition, 193

Principle of Multiple Root Causes, 193
Principle of Point of Control, 193
Prioritization process, 46, 49, 99, 104, 111, 153
Proactive monitoring, 152–153, 165
Problem-solving, 168–169, 182
Procedures, *see* Standard operating procedures (SOPs)
  implementation phase, 64, 66–68
  investigation questions, 204
Process control, importance of, 16–17
Process control instructions (PCI), 58, 66–67, 78, 122, 154
Process design:
  implications of, generally, 14
  risk assessment and, 82
  substandard, 218
  systems safety review, 60–61, 73
Process enhancement, factors in, 167–168
Process enhancement team (PET):
  behavioral safety management, 144
  defined, 14
  employee involvement, 167
  establishment of, 21
  implementation phase, 64
  inventory, role in, 139
  management commitment and, 46
  operational control, 113–114
  organizational chart, xxiii
  process failure/incident investigations, 190
  purpose of, generally, 227
  responsibilities of, xxii
  risk assessment, role in, 42, 83
  safety enhancement, 42
Process evaluations, 5
Process failure analysis, 5, 21, 169, 195
Process failure investigations:
  analysis, 190, 195, 199–200
  multiple root causes, 195, 197
  purpose of, 187–188
  questions for, 198–206
  required elements of, 189
Process flow, operational control, 113–114
Process flow diagrams, 14, 104, 144, 154, 183, 192, 226
Process improvements:
  design, 13–14
  implications of, generally, 42
  risk reduction strategies, 122–125
Process inputs:
  process risk analysis, 74–79
  uncontrolled, 143
Process management, 12–13, 30
Process Nonconformance/Accidents investigations:
  audit attributes, 256–257
  process policy, 197
  purpose of, 15, 49, 126, 188, 195
Process review, 8
Process risk assessment, 165
Process specifications, 13
Procurement, 168
  importance of, 61, 122, 129, 168
  investigation questions, 202
  review check sheet, 76–77
Product identification, traceability and, 118–119
Production processes, life cycle of, 17–19
Productivity, safety management and, 38–39
Product liability, 97
Product life cycle:
  planning and, 47
  risk assessment process, 40, 97–100, 219
Product/service design, EH&S considerations, 121–122
Profitability, 47, 219
Profitability, 47
Profit-sharing plans, 173
Progress reports, 208
Protectionists, 11
Protective equipment, 38, 123, 271
Psychological disorders, 37
Punishment, 173
Purchasing:
  audit attributes, 250
  contractors, management of, 120
  contract review, 119–120
  design controls, 121
  engineering controls and, 117–118, 121
  implementation phase, 218
  product identification and traceability, 118–119
  product/service design, EH&S considerations, 121–122
  suppliers, management of, 120
Pygmalion effect, ABC analysis, 160

Qualitative risk assessments, 81, 102, 104, 111
Quality:
  cost of, 6
  defined, 5
  measurement of, 5–6
Quality action teams, 14
Quality and environmental health and safety management systems (QEH&S):
  ISO 14000/9000 harmonized, 1–8
  planning considerations, 54–55

Quality and environmental health and safety management systems (QEH&S): (*Continued*)
  management system policy requirements, 51
  model objectives, 21–22
  principles of, 1–8
  program elements, 17–20
  risk, business and organizational, 81–83
Quality circles, 14, 167
Quality improvement committees, 194
Quality is Free (Crosby), 5
Quality Management Systems (QMSs):
  improvement goals, 9, 13–14
  process inputs, 11–12
  safety issues, 9, 27
Quality Management-TQM, safety program management *vs.*, 3
Quality manager, role of, 213
Quality system, 8, 232
Qualitative analysis, 177
Qualitative risk assessments, 81
Quantitative analysis, 177
Quantitative risk assessments, 81, 89–91

Raytheon, 22, 181
Reactive monitoring, 152–153, 165
Recordkeeping:
  audit attributes, 240–243
  implementation phase, 72–73, 218
  product/service design, 121
  training programs, 127
Redesign, 176
Red flags, 68
Redundancy, 6, 168, 183
Reengineering, 168, 197, 216
Refresher training, 128
Regulatory audits/review, 31, 54
Regulatory compliance, 58–59, 210
Reinforcement:
  importance of, 161–162, 169, 172–173, 217
  in safety management, 32–34
Reporting:
  internal, 49, 65
  standard procedures, 67
Reports:
  audit, 151, 212
  evaluation, 145
  management review, 213–214
  rejection/rework, 63
Reproductive illness, 37
Resource utilization, 30
Respirator programs, 38
Respiratory protection, 271
Retention, unfunded, 89, 93, 111
Return on investment, 133
Return-to-work programs, 60
Reward systems, 172–173
Rework, xxxvi, 4, 6, 41, 68, 168, 183, 189, 213–214
Right-to-know programs, 124
Risk acceptability, 105–106
Risk assessment:
  action plan, 111
  audit attributes, 235–236
  business and organizational QEH&S risk, 81–83
  business risk, identification of, 86
  employee-facilitated, 42
  implementation phase, 60
  importance of, 5–6, 13–16, 45
  measurement and evaluation, 137
  operational control and, 114
  process summary, *see* Risk assessment process
  risk financing, 88–95
  risk management strategies, 85–88
  risk reduction measures, cost-benefit analysis of, 83–85
  safety management and, 30–31, 41
  steps, generally, 86
Risk assessment process:
  acceptability of risk, 105–106
  categories, 106–108
  communications, 98–99, 107–108
  controls, identification of, 105
  documentation, 98
  exposure evaluation,104–105, 109
  hazards, identification of, 99
  initial review, 100–101
  mitigation process, 101–102
  MORT, 96–97
  objectives/targets, 100
  overview, 95–96
  people/processes, identification of, 104
  planning, 102–103
  product life cycle, 97–98
  risk ranking, 108–111
  risks, identification of, 98–99, 103–104
  summary, 95–111
Risk control:
  audit attributes, 244
  safety management and, 39–42
Risk identification:
  hazard, 99, 114–117
  in risk assessment process, 103–104
Risk management:
  importance of, 20, 217

integrated, 85–86
strategies, 85–88
Risk manager, role of, 50, 213
Risk Map (Microsoft), xxxiii, 20
Risk mapping, xxxiii, 89–93, 217
Risk prioritization, 82, 153
Risk reduction:
measures, cost-benefit analysis of, 83–85
progressive, 4
Risk transfer plan, integrated, 93–94, 105, 111
Root cause analysis, 21, 97

Sabotage, 15
Safety, importance of, 9
Safety committees, 14, 167, 194
Safety management:
action plans, 43
behavioral, 140–144
communications, 33–35
cost control, 39–42
effective programs, 32–33
indirect costs, 35–37
manager's role, 32
minor incidents to serious accidents ratio, 27
OSHA requirements, 26–29
productivity and, 38–39
risk control, 39–42
significance of, 27
supervisor's role, 32, 36
training, 33–35
Safety process enhancement groups, 141
Safety program management, 3
Safety Through Design, 45
SAIC, 287–288
Sampling, QEH&S, 146
Seagate Technologies, 181
Securities and Exchange Commission (SEC), 97
Self-monitoring, 162
Senior management:
communications, 88
inspections, role in, 138
integrated risk management process, role in, 93
training, 127
reviews, 207, 210–211. *See also* Management review
role of, generally, 19, 33, 42, 51, 53, 167, 217
training, 127
Service delivery, 119
Shareholder value, 40
Sick-leave, 40
Singhal, Dr., 21
Siting, 122
Six Sigma, 21, 23, 179–183
SMART goals/objectives, 46–47, 51, 53–54, 82, 162, 166, 168
Specifications, functional, 45
Stakeholders, influence of, 9, 48, 51, 53, 58, 61, 64–65, 71, 81, 123, 182, 209
Standard deviations, 90, 178
Standard operating procedures (SOPs):
benefits of, 2, 21, 49, 58, 66–67, 137
development of, 67
need for, 33
review of, 58
statistical process control (SPC) and, 176
Statistical process control (SPC), 91, 144, 174, 176–179, 181–182
Storage, 8, 121, 250–254
Strategic planning, business risks, 86
Stress:
investigation questions, 200–201
nonconformance and, 75
Subcontractors, 118
Substandard conditions, xxv, 32
Substandard performance:
causes of, 68, 137, 194
implications of, 188
Success factors, 219
Supervision, *see* Supervisors
guidelines for, 123–124
investigation questions, 204–205
review check sheet, 78–79, 146
Supervisors:
Behavioral Safety Technology (BST), role in, 141
empowerment of, 42
inventory, development of, 139
inspections, role in, 138–139
job-site inspections, 194
role of, generally, 7, 11, 14–15, 19, 36, 217
Suppliers:
as information resource, 63
management of, 120, 254–255
Surveys:
management role in, 5
safety and industrial hygiene, 4
Synergism, 197
Systematic risk, 2
Systematic risk assessment, 101
Systems safety:
engineering, 14
importance of, 30
programs, generally, 163
review, 4, 13, 60–61, 154
tools, 115–116

Tag-out procedures, 266–268
Target dates, 53, 139, 211
Targets, management commitment and, 53–55
Task analysis, 14
Teamwork:
  group-facilitated, 175
  process enhancement (PET), *see* Process enhancement team (PET)
  risk assessment, 102
  significance of, 136
Terminal symbol, in process flow diagrams, 183
Testing:
  audit attributes, 237–240
  control procedures, 151–152
  phase, components of, 45
Third-party audits, 114, 210, 217
Tools and equipment:
  audit attributes, 272–274
  implementation and, 218
  investigation questions, 205–206
  review check sheet, 79
Total quality environmental management (TQEM), 47
Total Quality Management (TQM), 2, 21, 29–30, 164, 197, 216
Trade-protection, 11
Training:
  administrative controls, 128
  audit attributes, 248–250
  commitment and policy, 2, 49, 52
  contractor personnel, 69
  emergency preparedness and response, 128–129
  hazardous material/waste spill response plans, 129–130
  implementation phase, 217
  importance of, 14, 19, 21, 219, 224–225
  management role in, 5
  medical surveillance programs, 130
  operational control, 125–130
  procedural controls, 125–126
  quality and environmental health and safety, 128
  safety management programs, 33–34, 39,128
  strategy, 127–128
Trauma disorders, 37
Trends, defined, 175

United Kingdom, safety management, 29–30
U.S. OSHA, 10, 13
Utilization cost, 219

Variable costs, 143
Vendors:
  certification, 61
  credentials, 118
  management and control of, 68–69, 218
  qualifications of, 118
Vision statement, 46

Waste management, *see* Hazardous materials/hazardous waste (HM/HW)
Waste reduction programs, 47, 99
Waste minimization, 8
Welch, Jack, 180
Wellness training, 39
Work center risks, 14
Work guidelines, 123
Workmen's compensation, 23, 41
Workplace hazards, 197
Work practices, evaluation of, 138
Work Risk Analysis and Control (WRAC), 116, 154

Zero defects, 22